Playa Systems

Edited by
Michael R. Rosen

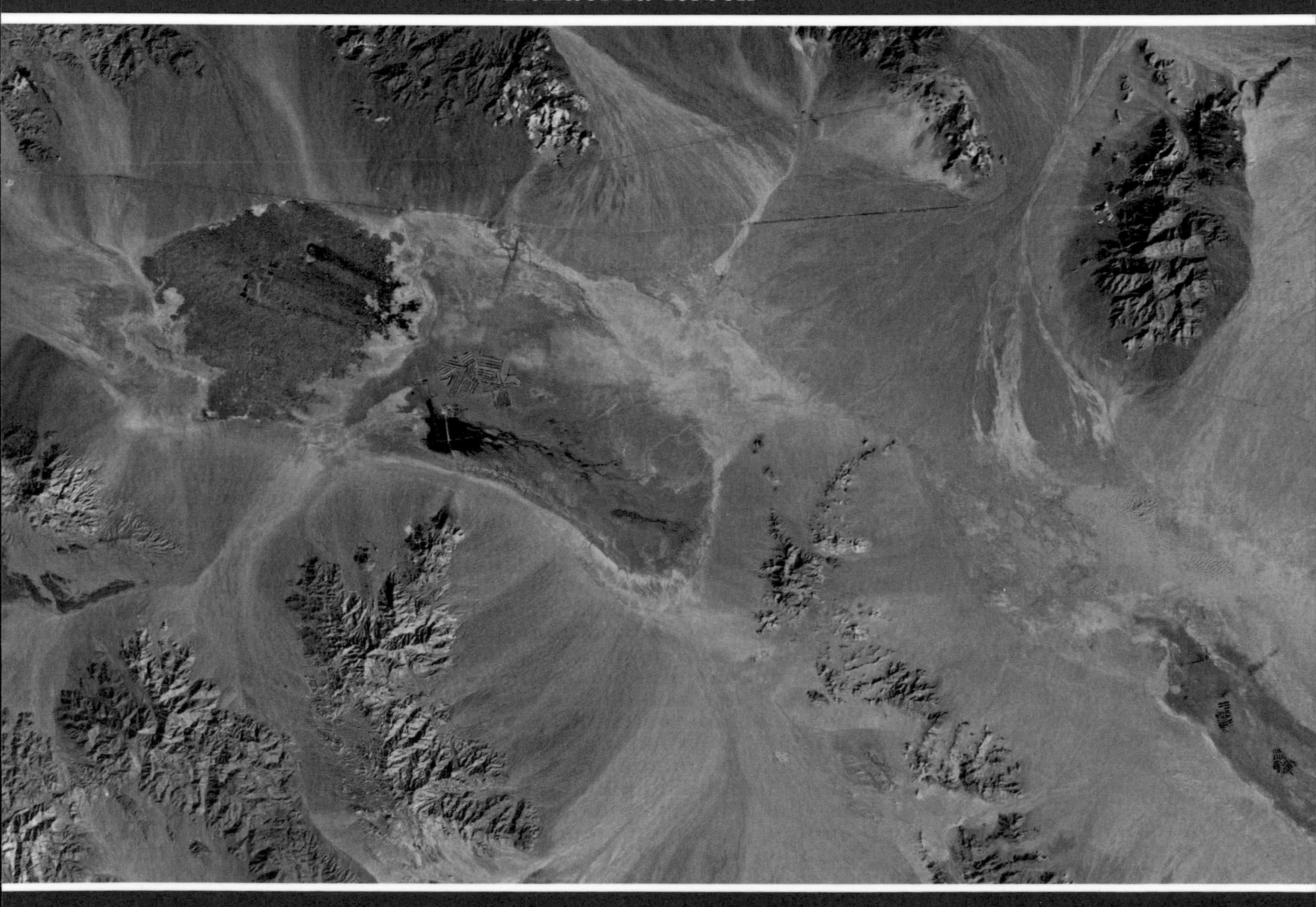

SPECIAL PAPER

289

Paleoclimate and Basin Evolution of Playa Systems

Edited by

Michael R. Rosen*
Limnological Research Center
University of Minnesota
220 Pillsbury Hall
310 Pillsbury Drive, N.E.
Minneapolis, Minnesota 55455-0219

289

*Present address: Institute of Geological and Nuclear Sciences, Wairakei Research Centre, Private Bag 2000, Taupo, New Zealand.

1994

Published by The Geological Society of America, Inc.
3300 Penrose Place, P.O. Box 9140, Boulder, Colorado 80301

Printed in U.S.A.

GSA Books Science Editor Richard A. Hoppin

Library of Congress Cataloging-in-Publication Data

Paleoclimate and basin evolution of playa systems / edited by Michael R. Rosen.
p. cm. — (Special paper ; 289)
Includes bibliographical references (p. –).
ISBN 0-8137-2289-6
1. Playas. 2. Paleohydrology. 3. Paleoclimatology. I. Rosen, Michael R., 1961– . II. Series: Special papers (Geological Society of America) ; 289.
GB612.P34 1994
551.4'15—dc20 93-50091
CIP

Front cover: Landsat Thematic Mapper (TM) image of the Bristol and Cadiz Dry Lake Basins. Bristol Dry Lake (left, 18 km across at its widest point) is highlighted by black, which represents standing water and halite crust in the basin center, surrounded by orange clays. Light blue around the margin of the playas represents loose sorted sand and yellow represents alluvial material. Cadiz Dune Field to the north of Cadiz playa (right) also appears as light blue. Pink colors may represent calcrete or gypsum deposits. Circular feature on the east of Bristol Dry Lake may be the remnants of a fan that reached farther out into the playa. The linear feature (running northwest-southeast) up near the fan in the same area is a low-lying shoreline barrier formed when there was standing water in the playa. Amboy Crater and lava flows (Pleistocene?) are prominent dark features to the left of Bristol Dry Lake and demonstrate how volcanics can influence the hydrology of the basin. The flows have cut off a small portion of the northwest section of Bristol Dry Lake from the main playa and formed Alkali Dry Lake. Black lines in the centers of both playas represent salt evaporation pans. TM image courtesy of the Jet Propulsion Laboratory, Pasadena, California. **Back cover:** *Left:* Ephemeral water dissolving the halite crust on Bristol Dry Lake, California. *Right:* Badwater Draw, a saline spring in Death Valley playa, is a relatively permanent groundwater discharge area in one of the driest regions of the world. *Center:* These halite tepee structures form due to groundwater discharging to the surface of the playa from a process that is almost identical to carbonate groundwater tepee formation.

10 9 8 7 6 5 4 3 2 1

Dedication

Kerry Kelts

This Special Paper is dedicated to Kerry Kelts, who is currently the director of the Limnological Research Center of the University of Minnesota. Kerry has been a primary influence in elevating the status of geological lacustrine research to a level of global co-operation. He has published work on saline basins in diverse areas such as Africa, Iran, China, and the United States, and on subjects such as lacustrine source-rock potentials, stable isotope paleohydrology, and paleoclimates. His early recognition that lacustrine sediments, in particular sediments in saline basins, could provide a key continental archive of paleoclimate records has provided much of the stimulus for the UNESCO-IUGS International Global Correlation Programme (IGCP) project 219, Comparative Lacustrine Sedimentology in Space and Time, and its successor project IGCP-324 Global Paleoenvironmental Archives in Lacustrine Systems. Kerry, a dedicated teacher and researcher, is truly an enthusiastic practitioner of his science.

Contents

Preface

The study of the sedimentology and geochemistry of lakes has received some attention in recent years because it has been realized that great amounts of information are stored in their sediments that can be used for paleoclimate and global change studies. Although there have been some significant advances in the studies of playas in recent years, there are very few collections of papers that are concerned only with playas.

This volume is a collection of papers that will help our understanding of both the basin dynamics and paleoclimate record in arid-zone lake systems. Playas are important because they may contain sedimentary records that are millions of years old, and in some cases they may have resolution on the order of tens of years or greater. These types of records are useful for both long- and short-term climate change studies. In this context, this volume is a contribution to the International Global Correlation Programme (IGCP) project 324, Global Paleoenvironmental Archives in Lacustrine Systems (GLOPALS). In addition, the water supply for irrigation or population centers in arid regions may greatly depend on the dynamics of groundwater associated with playa areas. A greater understanding of groundwater movement in playa zones may prevent overuse of these delicately balanced aquifers. In this context, the volume begins with a review that redefines the term *playa* and places it into a hydrological perspective.

The following papers deal with isotopic, sedimentologic, and hydrologic problems in diverse playa settings. The paper by Lowenstein et al. uses the fluid inclusions in halite to determine the Pleistocene/Holocene paleoclimate record from a terminal discharge playa complex in China. The results suggest that the regional aridity may have limited glaciation on the Qinghai-Tibet Plateau during the last glacial period.

The paper by Salama focuses on the little-studied buried lakes of the Sudanese rift system. Unlike the better-studied East African rift that contains saline lakes in the high uplifted areas, the Sudanese rift lakes are in the lowest portions of the basin where high sedimentation rates are encountered. Renaut and Last detail the sedimentology and geochemistry of lakes in two important areas of Canada, the Cariboo Plateau in British Columbia and the northern Great Plains of eastern Alberta, Saskatchewan, and southwestern Manitoba (Palliser's Triangle). These areas are very different in their geologic and geomorphic context, yet some of the same problems exist in both areas. Renaut details the sedimentology and brine evolution of Clinton Lake, a saline playa; yet he points out that the hydrology of the system is complex and requires further study. This situation is all too frequent in playa research. While Last focuses on a much larger geographic area, he too points out that much of the hydrological work remains to be done. He concludes that although the area is "ripe" for paleohydrological studies, because the area is in many ways unique, excessive comparisons with other playa regions may not be justified.

The remaining two papers of the volume highlight the importance of the Murray Basin area, which is a regional discharge area. In the first paper, Jacobson et al. describe a group of complex nested playas from the Mallee region of the Murray Basin. They illustrate the importance of distinguishing "fossil" brines from modern brines, and they demonstrate that brine pools migrate in space and time in response to changing climate.

The final paper in the volume by Talbot et al. compares the sedimentology of the Murray Basin playas with Triassic rocks in England. This paper illustrates that the study of playas is important not only to determine the basin evolution and paleoclimatic record of unlithified playa sediments, but also to identify criteria for the recognition of lithified, ancient playa sequences where parts of the basin morphology are missing.

This volume could not have been completed without the help of the reviewers, whom I would like to thank for contributing such an important part toward its completion. Technical support was provided while I was working at the Division of Water Resources, CSIRO in Perth Western Australia, and at the Limnological Research Center of the University of Minnesota. I would like to thank Kerry Kelts, director of the LRC, and Bruce Houghton, group manager of the Institute of Geological and Nuclear Sciences, for allowing me the time to complete this project.

Geological Society of America
Special Paper 289
1994

The importance of groundwater in playas: A review of playa classifications and the sedimentology and hydrology of playas

Michael R. Rosen*
Limnological Research Center, University of Minnesota, 220 Pillsbury Hall, 310 Pillsbury Drive, N.E., Minneapolis, Minnesota 55455-0219

ABSTRACT

Playas may form in both hydrologically open and closed systems. Several previous classification schemes of playas rely heavily on geomorphic features rather than the groundwater hydrology of the system and the degree to which the basin is hydrologically open or closed. This paper presents a review of the literature on playa classifications and their hydrologic characteristics, which can be used to better define the hydrology of playas. A modification of the original hydrologically based classification schemes of Meinzer (1922) and Snyder (1962) illustrates the importance of groundwater in playa systems.

In this paper a *playa* is defined as an intracontinental basin where the water balance of the lake (all sources of precipitation, surface-water flow, and groundwater flow minus evaporation and evapotranspiration) is negative for more than half the year, and the annual water balance is also negative. The playa surface must act as a local or regional discharge zone. Evidence of evaporite minerals will generally be present in parts of the basin. This includes carbonate minerals that can be demonstrated to have been formed through evaporative processes. A *recharge playa* is defined as above, except the playa surface acts as a means for recharging water to the aquifer. In this case, evaporite minerals are absent. Although evaporite formation is an important part of a playa sequence, significant accumulations of subaqueously deposited evaporites are only possible when saline water bodies are partially maintained by a constant inflow of groundwater.

In hydrologically closed basins, the brine chemistry is influenced by the lithology of the sediments and bedrock within the playa catchment. In hydrologically open (through-flow) discharge complexes, the brine composition may be determined by the rate of groundwater flow through the basin (leakage ratio) relative to the weathering rate of the surrounding sediments and bed rock. If the groundwater inflow greatly exceeds the rate of chemical weathering, then the brine composition may be dominated by the chemical composition of regional or local precipitation.

Extensive accumulations of subaqueous evaporites form in playas when there is sufficient groundwater to maintain a shallow brine pond for an extended period of

*Present address: Institute of Geological and Nuclear Sciences, Wairakei Research Centre, Private Bag 2000, Taupo, New Zealand.

Rosen, M. R., 1994, The importance of groundwater in playas: A review of playa classifications and the sedimentology and hydrology of playas, *in* Rosen, M. R., ed., Paleoclimate and Basin Evolution of Playa Systems: Boulder, Colorado, Geological Society of America Special Paper 289.

time. This has important implications for paleoclimate reconstructions of closed basin playa sequences. Extremely arid periods, when there is insufficient water to maintain a brine, will result in displacive intrasediment growth of evaporites; and extremely wet periods may be too fresh for a brine to develop. It is only in the intermediate periods when evaporation is high and water input is balanced by evaporation that extensive subaqueous evaporites will accumulate in closed basin playas.

INTRODUCTION

According to Neal (1975), there are approximately 50,000 playas on Earth. Although the individual area of playas is small, usually less than a few square kilometers, their study is important because of increasing population and agricultural pressure on the water supply in these regions, and the hydrology of the regions is becoming greatly stressed. Relatively little is known about the hydrology of many playa systems. Therefore, an increased understanding of playa hydrology and how irrigation and other anthropogenic activities may disrupt the natural balance of the system is important in order to solve these problems. Furthermore, there is a considerable literature on playa sedimentology and hydrochemistry, and yet, there is little communication between hydraulic considerations and geological investigation.

The purpose of this paper is to examine the literature on playa hydrology and sedimentology and to emphasize the importance of groundwater hydrology in the classification and recognition of playa systems. This will lead to a better understanding of the dependence of playa systems on groundwater discharge and may help to determine a more consistent classification of playa systems. It is also hoped that this paper will make geologists more aware of the hydrologic literature on playa basins so that their interpretations of playa basins are more consistent with how playa systems function. Similarly, it is hoped that hydrologists will become more aware of the sedimentologic aspects which pertain to the hydrology of the system. In most cases, the sedimentology and diagenetic history of a basin can be used as proxy hydrologic information in both modern and ancient playa settings.

Previous work. Throughout the century, there have been many attempts to classify playas and saline lakes (Meinzer, 1922, 1927; Clarke, 1924; Foshag, 1926; Thompson, 1929; Snyder, 1962; Motts, 1965; Reeves, 1968, 1978; Eakin et al., 1976; Hardie et al., 1978; Eugster and Hardie, 1978). Each of these classifications was based on a particular aspect or aspects of the basins to be defined. For example, Meinzer (1922), Snyder (1962), and Eakin et al. (1976) based their classifications on the hydrology of the Great Basin in the western United States, Eugster and Hardie (1978) and Hardie et al. (1978) classified saline lakes based on the evolution of the chemistry of the lake waters and inflowing groundwater, and Thompson (1929) and Reeves (1968) based their classifications on whether the playa was wet or dry. Foshag (1926) combined both the chemical aspects of playas with the geomorphic considerations of whether the playa surface was wet or dry.

This paper will review some of these classifications and demonstrate that the fundamental unifying element of these classifications is the groundwater hydrology of the systems in both hydrologically open and closed playa systems. Because the literature on playas contains many contradicting views on what constitutes a playa, some definitions are necessary.

Previous definitions of playas

When I first started studying evaporites, I was confronted by the term *playa,* which was used in the context of intermontane topographically closed basins in the southwest of the United States of America. At the time, this gave me a very narrow perspective of what I thought should be called a playa or playa lake. However, there are numerous nouns and adjectives used to describe evaporitic inland basins formed in widely varying hydrologic and geologic terrains. But what exactly is a playa or playa basin? How do the terms *continental sabkha, salina, dry lake, saline lake, ephemeral lake, salar, salt* or *clay pan, boinka, groundwater discharge zone* or *complex, playa lake,* and other terms relate to each other, and are there discrete definitions for these terms? Unfortunately, the study of inland saline basins, like many subdisciplines of geology, is mired in poorly defined or misused terminology, and this makes the meaning of all of the above terms confusing to the uninitiated.

The word *playa* means "beach" or "shore" in Spanish. Gilbert (1875, p. 86) seems to be the first to use the word in a geological context but he did not define it, and, although the word is italicized in his text, it is assumed that the reader knows the meaning of the word. This suggests that the word was known or defined from some previous text. He appears to be referring to waterlain deposits of ephemeral streams that consist of partially sorted gravels, saline clays, and sands.

Russell (1883) used both *playa* and *playa lake* in an article in a popular science magazine, but did not discuss the origin of the word. However, in an article published in 1885, he gives the best clue to the origin of the term. Russell described the mud-plains from ephemeral water-bodies which have evaporated to dryness as having wind-rippled mud deposits that "received markings that are usually considered characteristic of shores" (Russell, 1885, p. 82). He thus associated the Spanish word for shore to the surface of the basin sediments.

Russell (1885, p. 81) also defines a playa-lake as "inclosed water bodies of dry climates which have little depth and frequently evaporate to dryness, leaving mudplains or playas." This definition is in contrast to deposits of "permanent" saline water bodies that he would refer to as true lacustrine sediments.

According to Neal (1975, p. 1), *playa* is a "geological term for the flat and generally barren lower portions of arid basins of internal drainage that periodically flood and accumulate sediment." However, not all playas or playa lakes (flooded playas) have the same origin, and in many cases it can be demonstrated that the basins are not internally drained. Therefore, Neal (1975, p. 1) suggested that playa should then be used as a general term to describe a "variety of topographic depressions and desiccated former lakes that occur in the arid zone." This definition has been adopted by the geological community to the point that any desiccated basin regardless of origin has been termed a playa.

Other definitions of playa basins are based on how long the basin is wet or dry (Motts, 1965). If the basin is dry more than 75% of the time it is a playa, if it is wet more than 75% of the time it is a perennial lake. Basins that fall into the intermediate category are then called playa lakes (Motts, 1965).

However, local usage also confuses the terminology of arid-zone basins. In California, the term *dry lake* is used to describe the many intermontane basins found in the Basin and Range area (i.e., Bristol Dry Lake, Soda Dry Lake, etc.). The term is, in this context, synonymous with playa. In addition to playa, the terms *salina* and *sabkha* have also been used to describe inland evaporitic basins (see Warren and Kendall, 1985). Reeves (1968, p. 87) produced a fairly comprehensive list of local terms from various countries that are used to describe playas (Table 1). Interestingly enough, he did not include the Spanish use of *salina* in the list.

Unfortunately, many of these local terms have developed specific meaning in geological circles, making their use confusing and imprecise. An example of this is the term *sabkha* or *sebkha,* which means "salt marsh" in Arabic. Originally the term was applied to both coastal and inland saline depressions in North Africa and the Middle East (Christian et al., 1957). However, the term has been popularized by Curtis et al. (1963), Kinsman (1966), Butler (1969), Evans et al. (1969), and Warren and Kendall (1985) in a geological context as being restricted to marginal marine settings such as the Trucial Coast, with a defined stratigraphic sequence. Therefore, Handford's (1982a, b) use of the terms "Continental-sabkha playa basin" in describing Bristol Dry Lake in California is not incorrect (although redundant), but it is confusing in a geological context.

It could be argued that the playa is used to describe the center of the basin and that the continental sabkha describes the marginal mud-flat area surrounding the playa. However, it is clear from Gilbert (1875) and Russell's (1883, 1885) original definitions of a playa that this term refers to all of the low areas in the basin and not just the salt pan. Thus, Handford's (1982a) sabkha is easily identified as the *saline mud flat* subenvironment defined by Hardie et al. (1978). Furthermore, separating the basin into sabkha and playa reduces the nomenclature to describing the basin or geological sequence in terms of the individual facies and not the system as a whole. This leads to even more confusion in the literature.

TABLE 1. LIST OF LOCAL TERMS FOR EVAPORITIC BASINS*

Country/Region	Name
United States	Playa (dry), salina (wet), alkali lake, salt lake, playa lake, dry lake, salt pan, wetland (may be fresh)
Chile	Salar (much salt), salina (very little salt), tagarete (marshy area)
Southwest Africa	Vlor (dry lake), kalahari (salt lake), kalkpfannen (lime-encrusted lake)
North Africa	Chott, sebka or sebkha (dry lake or playa), merdja (playa)
Middle East	Sabkha or sabkaha (salt marsh, usually coastal), shott (playa)
Iran	Kewire (playa), dariache (an ephemeral desert lake)
Asia	Nor, sebehet, and schala (playa), tsaka, tsidam (salt lake or salt marsh)
India	Rei (playa)
Brazil	Praia
France	Plage
Spain	Salina, salada (salt lake or playa), laguna (small shallow lake but could be fresh)
Russia	Pljaž, pliazh (playa), takyre (dry lake)
Australia	Salt pan (playa), boinka (groundwater discharge complex), salina (coastal salt lake, usually wet)

*Modified from Reeves, 1968.

A thorough discussion of the sabkha nomenclature problem is given by Castens-Seidell (1984). She redefined the criteria used to recognize a sabkha deposit and contended that the term sabkha should be used only for marginal marine settings. In addition, she suggested that, because of the shoaling-upward nature of the marine sabkha setting, halite would not be part of a normal sabkha sediment package. This definition essentially precludes many hydrologically closed playas such as Bristol Dry Lake, and most other intermontane evaporitic lakes, from being termed sabkhas (Rosen, 1989).

Playas redefined. In this paper, a slightly more restricted approach than Neal (1975) to the term playa is used. For a depression to be considered a playa in this paper, all of the following conditions must be met. (1) The basin must be *intracontinental,* i.e., the regional shallow groundwater system does not directly discharge into the ocean. This excludes only those basins associated with coastal areas. This is because a separate nomenclature (and hydrology) exists for these depressions. Terms such as salina or sabkha would apply to these coastal basins. (2) The water balance of the lake (all sources of precipitation, surface water flow, and groundwater flow minus evaporation and evapotranspiration) must be a negative value for more than half the year, and the annual water balance should also generally be negative. This condition ensures at least periodic drying of the deepest part of the basin. (3) The capillary fringe is close enough to the surface of the playa such that evaporation will cause water to discharge to the surface. Discharge may be on a local or regional scale and does not imply that the basin is necessarily hydrologically closed. Hydrologically closed basins may be termed *discharge playas,* and playas that are known to have a regional output of groundwater (but are local discharge zones) may be termed *through-flow playas.* Evidence of evaporite minerals is usually present in some part of the basins, but usually in the lowest part. This includes carbonate minerals that can be demonstrated to have been formed through evaporative processes.

A special case of playas, here defined as a *recharge playa,* must meet the first two criteria listed above. However, in this case, the capillary fringe is far enough below the surface of the playa such that evaporation will not cause water to discharge to the surface. For fine-grained sediments such as clays, this means the water table will be greater than 2 to 3 m below the surface of the playa, but for coarser sediments the water table could be within a meter of the surface (Bear, 1979, p. 23). This condition implies that the playa is a site of groundwater recharge. Evaporite minerals will be absent from or a minor component of this type of playa. The Southern High Plains of Texas and New Mexico contain numerous examples of recharge playas (Osterkamp and Wood, 1987). A recharge playa is equivalent to the term *dry playa* of Thompson (1929).

The importance of evaporite minerals in defining a playa is related to the hydrological nature of the system and is discussed below.

SEDIMENTOLOGY

Geomorphic classifications of playas

The surfaces of closed basins can be flat or irregular, wet or dry, and these factors have been the basis for a number of classifications of playa types (Thompson, 1929; Reeves, 1968). Smooth surfaces may be moist or dry. The flatness of the surface is caused by deflation, which erodes the sediment down to the capillary zone where evaporating groundwater holds the minerals in place. This would lead to a smooth moist surface. Another way to develop a flat surface is to flood the playa, which dissolves the old crust and then deposits clastic mud on top. If the water table is sufficiently deep below the surface such that capillary zone does not effect the surface, then the surface of the basin may become hard-packed, smooth, and dry. This type of basin has been called a *dry playa* by Thompson (1929). Meinzer (1922) and Thompson (1929) recognized these types of valleys as areas where the groundwater table was far below the land surface. They suggested that this was because there was lateral leakage of the aquifer out of these topographically closed basins. Foshag (1926), using Thompson's suggestions, also postulated that these dry playas were caused by groundwater leakage or through-flow. Most importantly, he noted that dry playas do not accumulate saline minerals. Under my definition of a playa presented above, these dry playas would be considered *recharge playas.* This topic will be discussed further below.

Where weak seeps discharge into the playa itself or where evaporating water is drawn up to the playa surface, thick crusts of salts and carbonate form an irregular topography on the playa surface. Pustular algal mats associated with extrusive salt crusts and carbonates also form in the wetter areas of the playa irregularly dispersed over the playa surface. Extrusion of displacively growing salts can also occur around the margins of desiccation crusts and along tepee structures (lateral growth expansion features).

It is obvious, then, that the surface morphology of the playa surface is also controlled by the hydrology of the groundwater system under the playa. As the groundtable rises and falls due to availability of recharge, new seeps will develop and old ones will dry up. Similarly, as sediments accumulate in one area, water may pond in another. Over a significant amount of time, many different areas of the playa will have seen different hydrologic conditions. However, it should be stressed that the lateral facies (which are controlled by hydrology) may not be distributed vertically in the same sequence (see Kendall, 1988). The inherent anisotropy of the system should be expected because of the diversity of permeabilities associated with the lateral and vertical alteration of relatively insoluble siliciclastic mud and clay versus soluble (and permeable) saline minerals. This suggests that classifying playas by surface morphology can be difficult and perhaps misleading.

Prediction of a vertical facies model is complicated by the above features. Superimposed on this are diagenetic effects that may further alter the original textures (Rosen, 1991; Smoot and Lowenstein, 1991). Playa lake and saline lake stratigraphy should then be thought of in terms of a facies mosaic in which lateral and vertical hydrologic effects must be taken into account.

Handford (1981) proposed a geomorphic classification of characterizing sabkhas, including playas as part of a continuum of marine to continental-dominated processes. He suggested that, although many systems are a combination of all of the following, there are three end-member physical processes (excluding evaporation) that are most important: (1) marine, (2) fluvial-lacustrine, and (3) eolian-dominated conditions.

Aside from the nomenclature problems of using the term sabkha to describe continental systems (see above), this geomorphic classification is flawed in that one of its end members, the eolian-dominated class, does not rely on the physical properties of the wind to form. Rather, it is the presence of a shallow water table beneath the dunes, and at the surface in the interdunal depressions that creates an environment in which evaporites may form, and playas can develop. The reason that the wind is able to erode and deflate the basin is solely due to the fact that the sediment is not held in place by the capillary fringe of the water table. Thus, the basin will eventually erode to this equilibrium level. A further discussion of this concept related to the term *dry playa* is developed below.

Formation of evaporites in playas

The sedimentology and sedimentary structures of evaporite and nonevaporite minerals (dominantly siliciclastics) is beyond the scope of this paper, and has been dealt with in detail elsewhere (Hardie et al., 1978; Lowenstein and Hardie, 1985; Smoot, 1983, 1991). An excellent review by Smoot and Lowenstein (1991) summarized much of the modern literature for closed-basin sedimentology and briefly outlines some of the hydrological aspects of evaporite formation. However, the deposition of evaporite minerals must be discussed in this paper within the context of groundwater hydrology because the hydrology of the basin controls the mechanical and chemical deposition of evaporites within the basin.

Saline minerals can form in all parts of the basin and recharge area, although the ultimate significant accumulations of saline minerals will be in the deepest part of the basin (Fig. 1). Saline minerals form by the following types of chemical and mechanical concentration: (1) Surface efflorescent crusts form either as the final stage of evaporation of a standing body of water or by evaporation of capillary zone or groundwater discharge on, or near the playa surface. The mineralogy and topography of the playa surface reflect the hydrology of the discharge area and can be morphologically complex. (2) Subsurface intrasedimentary growth of saline minerals results from precipitation from interstitial brines (groundwater). Extensive evaporative concentration results in displacive and

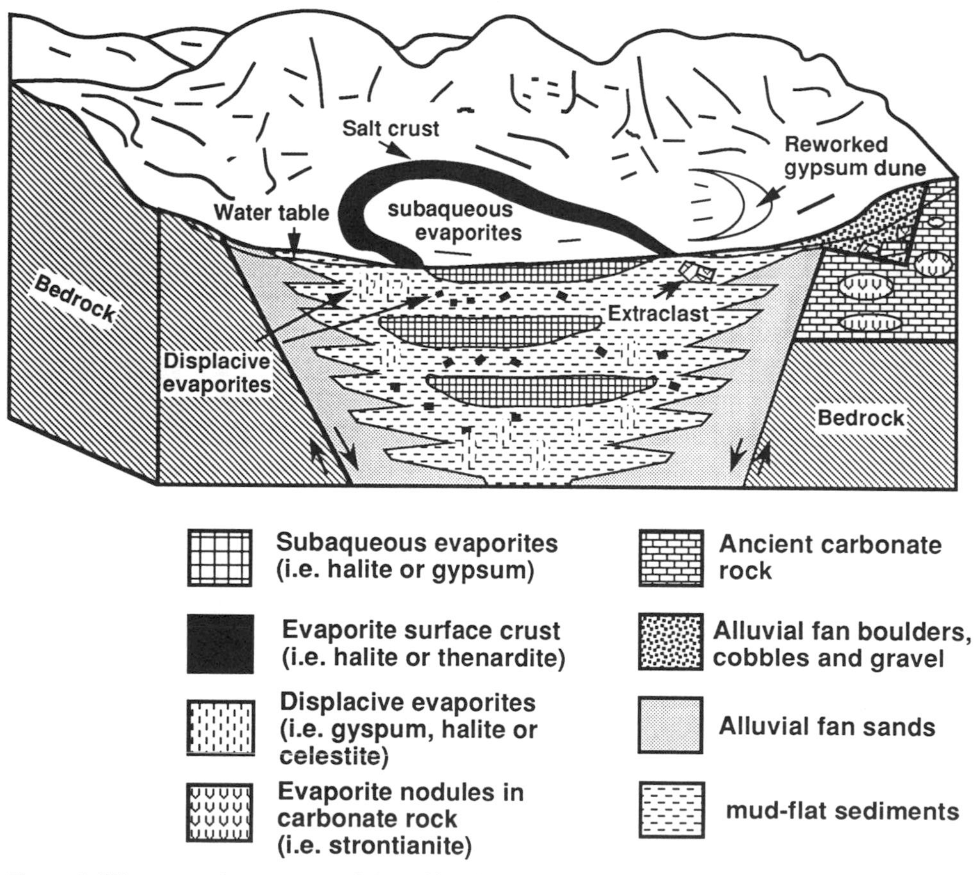

Figure 1. Diagrammatic summary of depositional environments of evaporite minerals (not to scale). Salts derived from primary evaporative concentration of brines include (1) subaqueous evaporites, (2) crusts, and (3) displacive evaporites. Secondary deposits derived from the redeposition of existing evaporites include (1) intraclasts (derived from erosion of evaporites precipitated in the playa), and (2) extraclasts (derived from evaporites derived from the surrounding bedrock). Salt crust may form across the entire playa floor when there is no standing water in the basin. Gypsum dune is derived from the deflation of gypsum crystals from the playa surface and may be considered as an evaporite intraclast deposit.

poikilitic growth that may concentrate into layers or nodules. (3) Single crystals or aggregates may form on the surface of a standing brine body (i.e., halite rafts), or at any depth in the water column, and settle to the bottom. Subaqueous growth of selenite crystals (gypsum >1 cm) on the substrate up into the water column has been documented in coastal salinas (Warren, 1982a, b; Castens-Seidell, 1984), but not in Basin and Range type playa lakes. (4) Clasts of saline minerals deposited in one part of the basin and redeposited in another part (intraclasts) represent a mechanical means of accumulating saline minerals. Detrital input, usually carbonates, from the surrounding highlands (extraclasts) may also be important (i.e., Deep Springs Lake, California). (5) Deflation of the playa surface during relatively dry periods may result in the formation of gypsum and/or clay pellet dunes (i.e., Salt Flat Playa, West Texas, and Willandra Lakes system, Australia). Although these dunes may not be as potentially preservable as the other methods of saline accumulation, stabilization by plant cover may preserve some dune morphology.

Of the three chemical methods for evaporite formation listed above, two are directly dependent on the discharge of groundwater into the basin (1 and 2), and the third may be caused by a high rate of groundwater discharge, concentration of surface runoff, or a combination of the two.

However, for a significant accumulation of subaqueous halite or gypsum to occur in a closed basin, there is likely to be a significant proportion of groundwater entering the lake. This is because if groundwater is not important, then a completely runoff-fed lake will not last long and will evaporate before significant evaporite minerals could accumulate. Only repeated runoff events over a period of time would keep the basin full enough to produce a significant accumulation of evaporites. But if there are repeated runoff events discharging into the basin, then the groundwater table must rise and eventually produce a permanent lake. For the lake to achieve a balance of greater evaporation than recharge, such that the water body is permanent, the lake level must be groundwater controlled.

Presumably, this balance of net evaporation minus recharge, if large enough, will produce a saline lake and potentially a significant accumulation of subaqueously precipitated evaporite minerals. In addition, if surface runoff was the sole contributing factor in evaporite precipitation, then to produce even a 3-m-thick accumulation of halite, over 200 m of freshwater would be required (Holser, 1979). However, most inland basins can be demonstrated to have been much shallower when halite was being precipitated (Rosen, 1991). Instead of single runoff events, a somewhat steady addition of water at a rate slower than evaporative concentration would maintain a permanent water body which, at a much shallower lake level, would produce significant accumulations of evaporite minerals (Fig. 2A to D).

In essence, what this means is that playas cannot accumulate significant amounts of subaqueous evaporites because they are usually dry or do not have a sufficient volume of

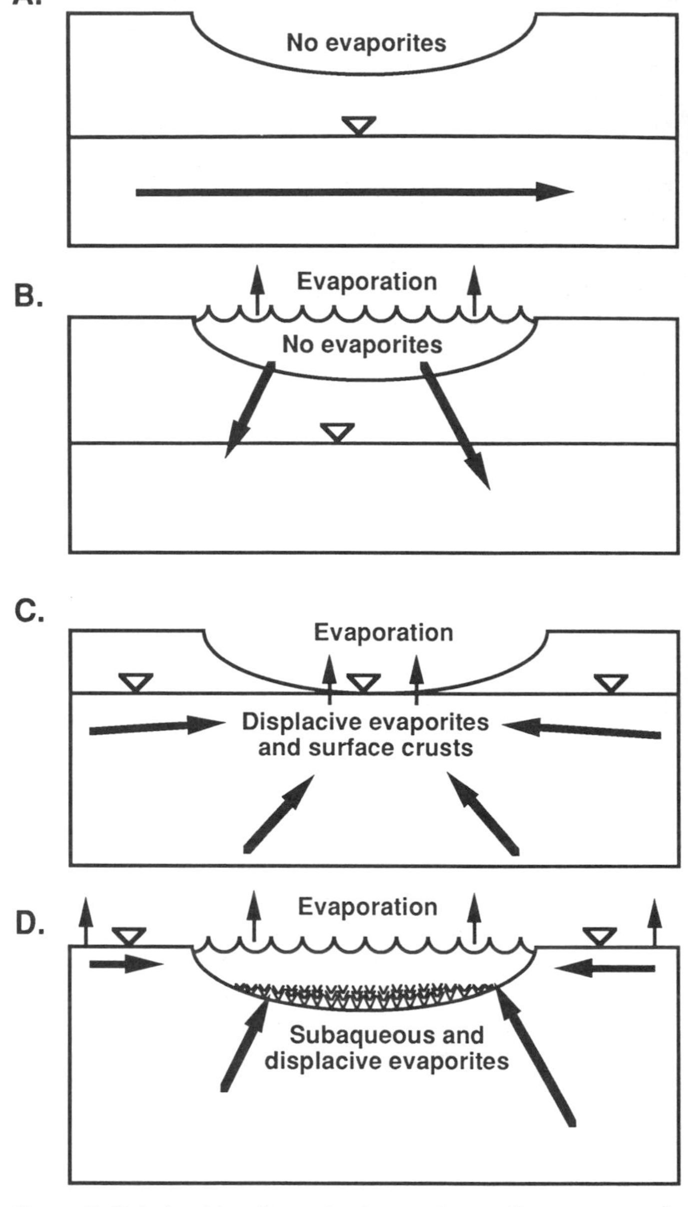

Figure 2. Relationship of sustained groundwater flow to evaporite mineral accumulation to all possible groundwater configurations. When the water table is far below the ground surface only flow through conditions can occur and no evaporites will accumulate (A). When water is transported or precipitated in this type of situation, the depression acts as a recharge zone and leaks water to the subsurface (B). No significant evaporites will accumulate although a thin crust may develop when the final solution evaporates to dryness. This crust would likely be deflated. When the water table intersects the ground surface in a hydrologically closed basin (a playa situation), displacive evaporites may form, but significant accumulations of subaqueous evaporites cannot (C). Those that can accumulate after a rain event will likely be deflated when the lake dries out. However, significant accumulations of displace evaporites may occur. Only when the groundwater table is above the surface of the deepest part of a closed basin playa, so that groundwater input is constant (D), can subaqueous evaporites accumulate in a hydrologically closed basin. Although slightly more complicated for through-flow basins, this model also applies to these types of basins.

water to produce thick evaporite beds. Only permanent saline lakes maintained by some component of groundwater flow can form thick evaporite beds. A permanent lake fed by surface flow must have some component of groundwater input or else the lake would drain (Fig. 2B).

For example, Bristol Dry Lake in California has arguably been a playa for at least the last 10,000 years, yet the amount of halite accumulated on the surface is only about 40 cm (Rosen, 1989). Furthermore, from historical records it can be demonstrated that the halite is only about 12 years old. Thus, as a playa, very little halite has accumulated in the basin center even though repeated large (but short term) runoff events have contributed large volumes of freshwater to the basin. However, in long cores taken in the center of the basin, there are thick (>10 m), pure beds of halite (Rosen, 1991). These halite beds must represent a time when the playa was in a permanent-lake phase, with a more constant supply of water entering the basin at least partially via groundwater while maintaining a highly saline permanent lake.

This has important implications for paleoclimate reconstructions of playa sequences. Extremely arid periods, when there is insufficient groundwater to maintain a brine, will not result in significant subaqueous evaporite accumulation, and extremely wet periods may be too fresh for a highly concentrated brine to develop. It is only in the intermediate periods when evaporation is high and groundwater input is sufficient that subaqueous evaporites will accumulate in hydrologically *closed* basins.

However, playas do accumulate significant amounts of groundwater-derived interstitial evaporites and even though there are no surface evaporites accumulating, diagenetic processes are important. In addition, playa basins cannot be static or else no sediment or subaqueously precipitated evaporites will accumulate. Therefore, although subaqueous evaporites accumulate in permanent saline water bodies, they are still part of an overall playa sequence and stratigraphy if the basin can be shown to periodically desiccate.

In summary, the chemical formation of evaporites will usually be volumetrically more important than the mechanical deposits (4 and 5), and because the chemical deposits are usually in a deeper part of the basin, will be, in most cases, more preservable. Therefore, the accumulation of evaporite minerals in playa basins is dependent on the hydrology of the system and the recognition of fabrics produced by different hydrological regimes is important in determining the evolution of the basin (see Teller and Last, 1990; Rosen and Warren, 1990; Smoot and Lowenstein, 1991).

HYDROLOGY

Hydrologic classification of playas

The only real hydrologic classification of playas was first proposed by Meinzer (1922). This scheme was then elaborated on by Snyder (1962) and then modified by Motts (1965), Eakin et al. (1976), and Mabbutt (1977). Snyder's (1962) classification scheme incorporated both topography and playa type, but, because he was concerned exclusively with the hydrology of the Basin and Range and the Great Basin of the western United States, his classification scheme was incomplete. However, it should be noted that he was really attempting to classify *valley types* in the western United States for water-use purposes and he was not primarily concerned with classifying different types of playas.

Hydrology of playas

Hardie et al. (1978) stated that to form a saline or playa lake basin, two conditions must be met. First, evaporation must exceed inflow, and second, the basin should be hydrologically closed or greatly restricted. Restriction of the basin may be caused by many things. However, the most favorable conditions for the formation of saline lakes are in rain-shadow basins. Such basins provide the high relief around the basin necessary to trap precipitation, while the basin floor remains arid (Eugster and Hardie, 1978). However, there are many other ways to create what many people have called playas. In South and Western Australia, for example, many of the so-called playas are the result of local or regional discharge of a through-flowing groundwater system (Macumber, 1991; Commander et al., 1991; Turner et al., 1992). These types of systems have been termed *boinkas* by Macumber (1991, and references therein). Unlike playas, which result from rain-shadow areas that have high relief, the relief of the drainage area surrounding Australian boinkas is low. Furthermore, although evaporation greatly exceeds precipitation, these basins are not hydrologically closed or even greatly restricted, rather they are the local surface discharge of a regionally extensive groundwater system. Similar types of local discharge zones in regional aquifers exist in West Texas (Wood and Sanford, 1990) and the Great Plains area (Winter and Carr, 1980; Arndt and Richardson, 1993) of the United States. This subject will be more fully discussed below.

Water sources. The water of a playa brine can be derived from two main sources: (1) groundwater and (2) direct precipitation as either rain or snow in the catchment area (including rivers and all surface inflow). Liberation of fluid from fluid inclusions of weathered detrital grains may be responsible for an extremely small (<1%) volume of new water entering the groundwater. However, there are many different components to both direct precipitation and groundwater inflow. Direct precipitation may be from a regional weather pattern, falling as rain or snow within the catchment, or as river or ephemeral stream input resulting from overflow from another basin.

Groundwater may be derived from (1) the local or regional meteoric groundwater system, (2) connate or formation water trapped in sediments (which may have a marine origin), or (3) deep-basinal or extra-basinal hydrothermal fluids (Fig. 3). Because many large playa systems occur in tectonically ac-

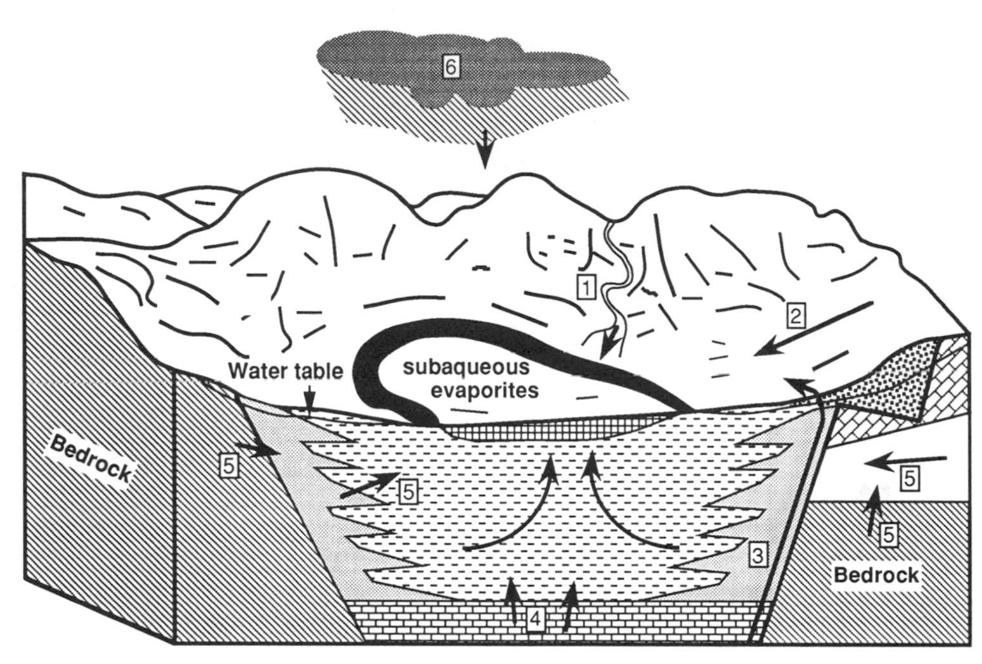

Figure 3. Sources of groundwater to a playa basin: 1, Channeled flow in permanent or ephemeral streams; 2, unrestricted overland flow (sheet wash); 3, hydrothermal fluids from a deep source (may occur as a seep or spring or directly mix into the groundwater); 4, connate or formation water derived from when the formation was deposited; 5, meteoric groundwater derived within (hydrologically closed) or outside (through-flow) the immediate basin; 6, direct precipitation onto the playa surface or surrounding catchment.

tive areas, the hydrothermal component of groundwater input, particularly in terms of the origin of solutes to the basin, may be very important (Hardie, 1984, 1990). For example, Lowenstein et al. (1989) were able to show that even small volumes (>2.5%) of hydrothermal water may significantly change the chemical composition of playa brines. However, Rosen (1991) was able to demonstrate that a conservative element such as chloride can be accounted for by constant input from precipitation and a slow sedimentation rate, without the need to invoke hydrothermal water.

Jankowski and Jacobson (1989) and Jacobson and Jankowski (1989) also showed that the amount of rainfall on a given part of a catchment area may be important in determining the chemical evolution of the brine. In the Lake Amadeus region of central Australia, the higher rainfall areas have brines that become bicarbonate dominated and the areas of lighter rainfall evolve to chloride-dominated brines. The difference in the brine compositions is related to dissolution of the carbonate sediments within the catchment. Where the rainfall is high, more carbonate can be dissolved and contribute to the brine composition, although the higher rainfall areas would also have a shorter groundwater residence time and so a shorter period of time for water-rock interaction than the low rainfall areas (Jacobson et al., 1991).

Water chemistry of playas. Lithology of the hydrologic basin surrounding a playa plays an important role in defining the configuration and distribution of flow system boundaries and the chemistry of the resulting playa brine. For example, in an area where there is sufficient precipitation, a high hydraulic gradient may be established in low-permeability rocks. High hydraulic gradients result in an intersection of the zone of saturation with the land surface, and local discharge may develop and form perennial streams, seeps, or springs. Thus, the presence of tufa in the Wilkins Peak Member of the Eocene Green River Formation (Eugster and Hardie, 1975; Smoot, 1983) may indicate a relatively high hydraulic gradient in a relatively low permeability setting or it may simply indicate the proximity of the playa and the zone of discharge. If the seep developed is of sufficient volume in a low topographic area, a boinka (discharge complex) may develop.

Although lithology is important, the degree to which the system is open and the amount of rock-water interaction in relation to the volume and flow-rate of the aquifer are also important characteristics of the chemical make up of playa basins. For example, in North America and Africa it can be shown from numerous basins that lithology is dominant in determining the chemical makeup of the playa water (Jones, 1966; Jones et al., 1977; Hardie et al., 1978; Eugster and Hardie, 1978). However, in Australia this is not necessarily the case. It has been clearly documented by numerous authors that with few exceptions, even the most inland playas are dominated by water chemistries which are ultimately derived from seawater regardless of the surrounding lithologies (McArthur et al., 1989; Herczeg and Lyons, 1991; Chivas et al., 1991).

Wood and Sanford (1990) and Sanford and Wood (1991) have studied brine compositions by developing a lumped-parameter model that calculates the thermodynamic geochemical saturation state of the brine. They determined that the leakage ratio (ratio of water outflow to inflow), which is also equivalent to through-flow, will have a profound effect on both the evaporite mineral suites precipitated and the thickness (or volume) of the deposit. Even small leakage ratios will greatly affect the evaporite mineral assemblages.

In essence, leakage (or through-flow of water) will stop the brine from evolving to higher concentrations and will eventually cause the brine to reach an equilibrium concentration beyond which it cannot evaporate. Thus, thick sequences of single evaporite minerals may develop in hydrologically open basins that cannot be explained by the lithology of the basin. This model demonstrates the importance of groundwater movement in controlling evaporite depositional sequences.

Origin of the chemical composition of brines. In the past, emphasis has been placed on geochemical parameters and hydrologic systems in inland evaporite settings in which the variation in evaporite mineralogy in a given basin can be constrained by fluid flow pathways and the availability of ions from the parent solution (Hardie, 1968; Hardie and Eugster, 1970; Hardie et al., 1978; Eugster and Hardie, 1978; Smith et al., 1987; Schmid, 1988). The evolution and chemical composition of brine solutions in closed-basin settings are probably the best-studied aspects of saline lakes. Initially, Jones (1966) discussed the evolution of closed-basin brines within the context of Western Great Basin (Basin and Range Province) saline lakes. Jones and Van Denburgh (1966), Hardie and Eugster (1970), and Eugster and Hardie (1978) expanded the basic premise of Jones (1966) and applied it to all closed-basin systems. Hardie et al. (1978) summarized previous work while emphasizing a sedimentologic framework.

Jones (1966) was the first to recognize that the composition of saline water in closed, inland basins is inherited from the chemical weathering processes of the drainage basin. In other words, the chemistry of the fluids coming into the basin is a product of the original mineralogy of the surrounding rocks. However, as mentioned above, significant quantities of solutes can be derived from the original rain water input into the basin. Evaporation of the incoming water in the basin leads to selective mineral precipitation. In simple terms, the genesis of saline waters was thought to be a function of solute supply and evaporative concentration. This is probably true for hydrologically closed systems. However, most basins are to some degree hydrologically open. Wood and Sanford (1990) have shown that the genesis of saline waters, in this situation, may be controlled more by the leakage ratio (see above) than by the amount of evaporation. The evolution of the water can then be described by solution, transport, and mineral precipitation reactions in sequence; and this sequence is controlled by basin hydrology and the degree to which the basin is hydrologically open.

In a hydrologically closed basin, primary solute composition is determined mainly from reaction of natural waters with lithologies underlying and surrounding the drainage basin. However, secondary modifications, such as the mixing of two chemically different inflow waters, to the solute composition are controlled by hydrologic setting and process. Classification of brine type (Table 2, Fig. 4) was based on recognition of major anions in solution because they were thought to have a more diverse origin than the cations (see Jones, 1966; Hardie and Eugster, 1970).

In the absence of older chemically precipitated deposits, water compositions are a product of (1) silicate hydrolysis, (2) uptake of CO_2 from the atmosphere and/or sulfate from oxidized sulfides, (3) precipitation of alkaline earth compounds, and (4) the original contribution of solutes from rain water and aerosols. If older chemical precipitates (evaporites or carbonates) or clays are present in the drainage basin, simple solution

TABLE 2. MAJOR EVAPORITE MINERALS OF THE DIFFERENT BRINE TYPES IN FIGURE 4*

Brine Type	Saline Mineral	
Ca-Mg-Na-(K)-Cl	Antarcticite	$CaCl_2.6H_2O$
	Bischofite	$MgCl_2.6H_2O$
	Carnallite	$KCl.MgCl_2.6H_2O$
	Halite	NaCl
	Sylvite	KCl
Na-(Ca)-SO_4-Cl	Tachyhydrite	$CaCl_2.2MgCl_2.12H_2O$
	Gypsum	$CaSO_4.2H_2O$
	Glauberite	$CaSO_4.NaSO_4$
	Halite	NaCl
	Mirabilite	$NaSO_4.10H_2O$
	Thenardite	$NaSO_4$
Mg-Na-(Ca)-SO_4-Cl	Bischofite	$MgCl_2.6H_2O$
	Bloedite	$Na_2SO_4.MgSO_4.4H_2O$
	Epsomite	$MgSO_4.7H_2O$
	Glauberite	$CaSO_4.NaSO_4$
	Gypsum	$CaSO_4.2H_2O$
	Halite	NaCl
	Hexahydrite	$MgSO_4.6H_2O$
	Kieserite	$MgSO_4.H_2O$
	Mirabilite	$NaSO_4.10H_2O$
	Thenardite	$NaSO_4$
Na-CO_3-Cl	Halite	NaCl
	Nahcolite	$NaHCO_3$
	Natron	$Na_2CO_3.10H_2O$
	Thermonatrite	$Na_2CO_3.H_2O$
	Trona	$NaHCO_3.Na_2CO_3.2H_2O$
Na-CO_3-SO_4-Cl	Burkeite	$Na_2CO_3.2Na_2SO_4$
	Halite	NaCl
	Mirabilite	$NaSO_4.10H_2O$
	Nahcolite	$NaHCO_3$
	Natron	$Na_2CO_3.10H_2O$
	Thenardite	$NaSO_4$
	Thermonatrite	$Na_2CO_3.H_2O$

*After Eugster and Hardie, 1978.

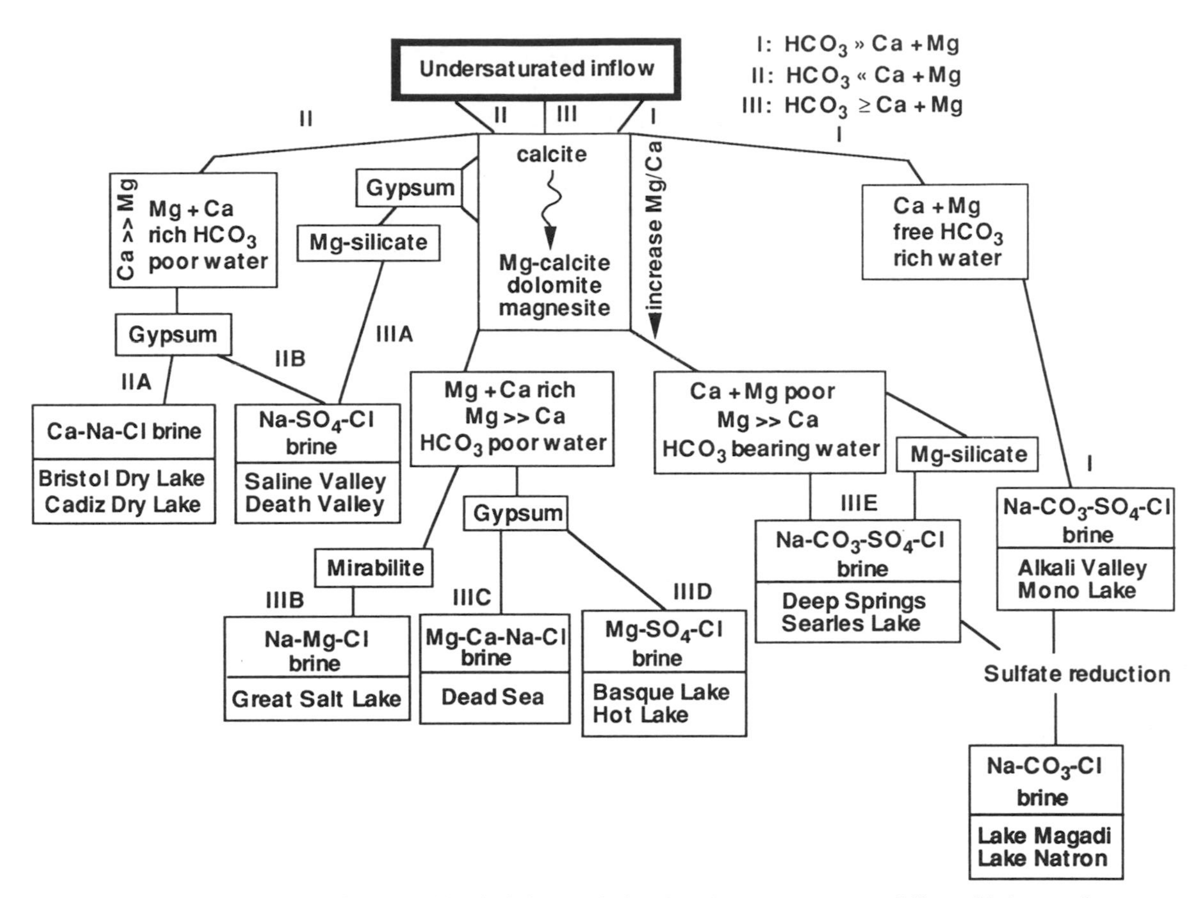

Figure 4. Flow chart for closed basin brine evolution based on computer modeling of brines and field examples (after Eugster and Hardie, 1978).

and leaching of adsorbed ions may contribute to the initial brine composition (Eugster and Hardie, 1978; Hardie et al., 1978).

The work of Wood and Sanford (1990) and Sanford and Wood (1991) has shown that the initial divides for brine types in the Eugster-Hardie scheme may depend more on the leakage ratio rather than the lithology of the surrounding basin. Only those basins that can be shown to be completely hydrologically closed will follow the paths in Figure 4 to completion. However, there are many more hydrologically open basins than closed ones, and it could be argued that almost all basins leak to some degree.

For example, the water chemistry of most playas in Australia can be shown to have been derived from seawater. This is probably due to the fact that most Australian playas are to some degree hydrologically open. The schemes of Jones (1966) and Hardie and Eugster (1970) described above were modeled on systems which are, for the most part, hydrologically closed. In a closed system, the residence time of the water in contact with a given lithology is longer than in an open system. Therefore, in an open system, the rate of solute input from precipitation and outflow through leakage may be faster than the rate of weathering of the surrounding bedrock. Due to the open system, the rate of removal of solutes derived from weathering is faster than they can accumulate in the groundwater. This leads to a playa where the water chemistry is controlled by the leakage from the system and the composition of the precipitation entering the system.

Conversely, in a closed system, where the groundwater has a long residence time and can accumulate in the groundwater at a faster rate than the solutes derived from precipitation, the playa chemistry may be dominated by the surrounding lithology.

Of course, many apparent exceptions may occur due to hydrothermal activity or the local intricacy of the groundwater system, and it should be noted that different solutes may have different origins, or the same solute may have more than one origin. However, in general terms, the derivation of the solutes in a system is related to the degree to which the system is open or closed.

Modeling of playa systems. Hydrologic computer modeling of closed basin playa systems is in an early stage, mostly because many of the parameters necessary to adequately model the systems have only recently been quantitatively studied. Although models for through-flow lakes (hydrologically open systems) are plentiful (Winter, 1976, 1978; Townley and

Davidson, 1988; Townley and Turner, 1992; Townley et al., 1992; Wood and Osterkamp, 1987; Wood and Sanford, 1990; Turner et al., 1992; Donovan and Rose, 1992; Ferguson et al., 1992) and are applicable to open-system playas, there have been relatively few attempts to model hydrologically closed basins. In fact, many of the above models were used to show that although evaporites occur in many basins, the systems studied are hydrologically open (Wood and Osterkamp, 1987; Wood and Sanford, 1990; Turner et al., 1992; Donovan and Rose, 1992).

There are, however, only a few studies on closed basin modeling, and in fact, this may be because closed basins are easier to physically model than through-flow basins (L. Townley, written communication, 1993).

Langbein (1961) was probably the first to theoretically model the salt content and hydrology of closed basin systems. In this important paper, he showed that the mass balance of salt inputs and losses could be modeled if geometric and hydrologic factors in the basin were known. He then derived theoretical equations for estimating the mass balance of salts from these parameters. Many of the subsequent computer mass balance models for both closed and leaky lake basins are based on this work.

Mifflin (1968) introduced the concept of *terrane flow capacity* as the amount of water any particular hydrologic environment could accept and transmit. This concept has been defined as the *terrane capacity* (Q_t) and is directly related to Darcy's Law. The regional parameters that control Q_t are average slope of terrain and average permeability. Should enough moisture for maximum recharge be available, these two variables determine the amount of transmittable groundwater before rejection by discharge from the terrain. Figure 5A to C illustrates Q_t for three situations typical in the Basin and Range. This concept may be helpful in determining geochemical sources or gradients for observed precipitated mineral sequences and brine compositions. In other words, does the composition of a particular measured brine reflect major or minor interaction with the host rock or alluvial sediments?

Duffy and Al-Hassan (1988) used empirical data and computer models to simulate the groundwater flow of the well-studied Great Basin (western United States). This area is probably one of the best studied hydrologically closed basin systems in the world, and is the area from which most of the early classifications of playas were developed.

Duffy and Al-Hassan (1988) determined by computer simulations that the relief of the surrounding mountains and the Rayleigh number (the ratio of buoyancy forces tending to cause flow to other forces tending to resist flow; i.e., free convection) will determine the hydrochemical cycling in undrained closed basins (Fig. 6A to D). For a hydrologically closed basin with high relief, and thus a high rate of upland recharge, the Rayleigh number would be low because a high recharge velocity decreases the Rayleigh number and so retards free convection (Duffy and Al-Hassan, 1988). On the other hand, a basin with low relief and low precipitation rates will have a high Rayleigh number and so promote free convection.

The model then predicts that this type of convection would promote degradation of the freshwater–saline water contact and eventually make the entire basin saline. This is exactly what is seen in the salinity patterns of large groundwater controlled playas such as Bristol Dry Lake, California. In ad-

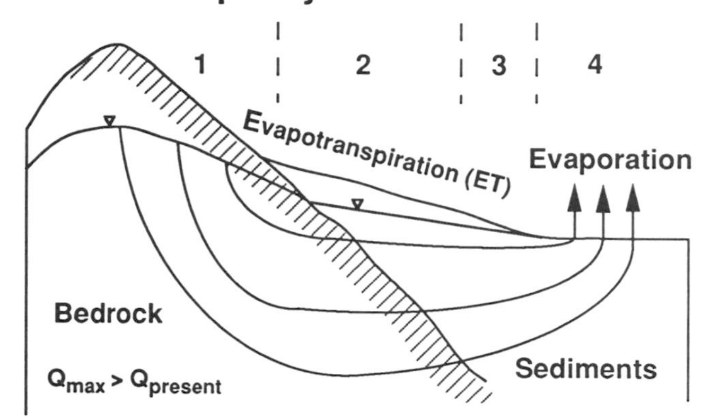

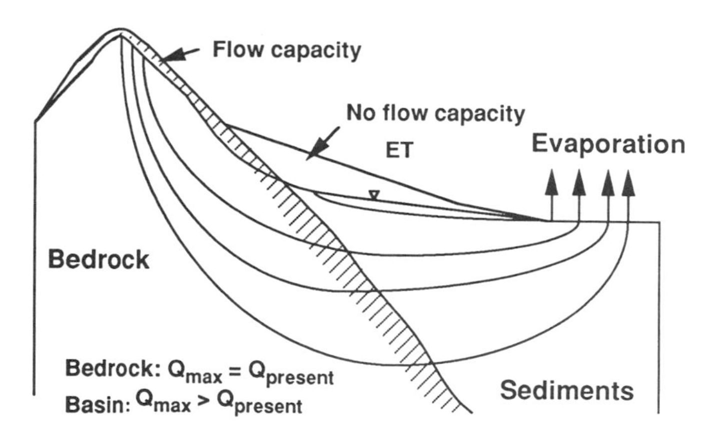

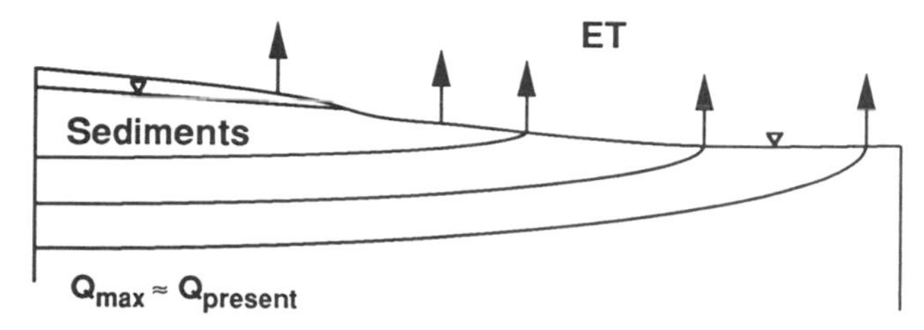

Figure 5. Diagrammatic sketches of the concept of terrain capacity for regional groundwater flow (after Mifflin, 1968). A, no flow capacity situation; B, flow capacity in bedrock present but not in fan; C, flow capacity in alluvial fan. Numbers in Figure 5A refer to (1) recharge area, (2) area of through-flow (lateral flow), (3) area of increased groundwater potential, and (4) discharge area.

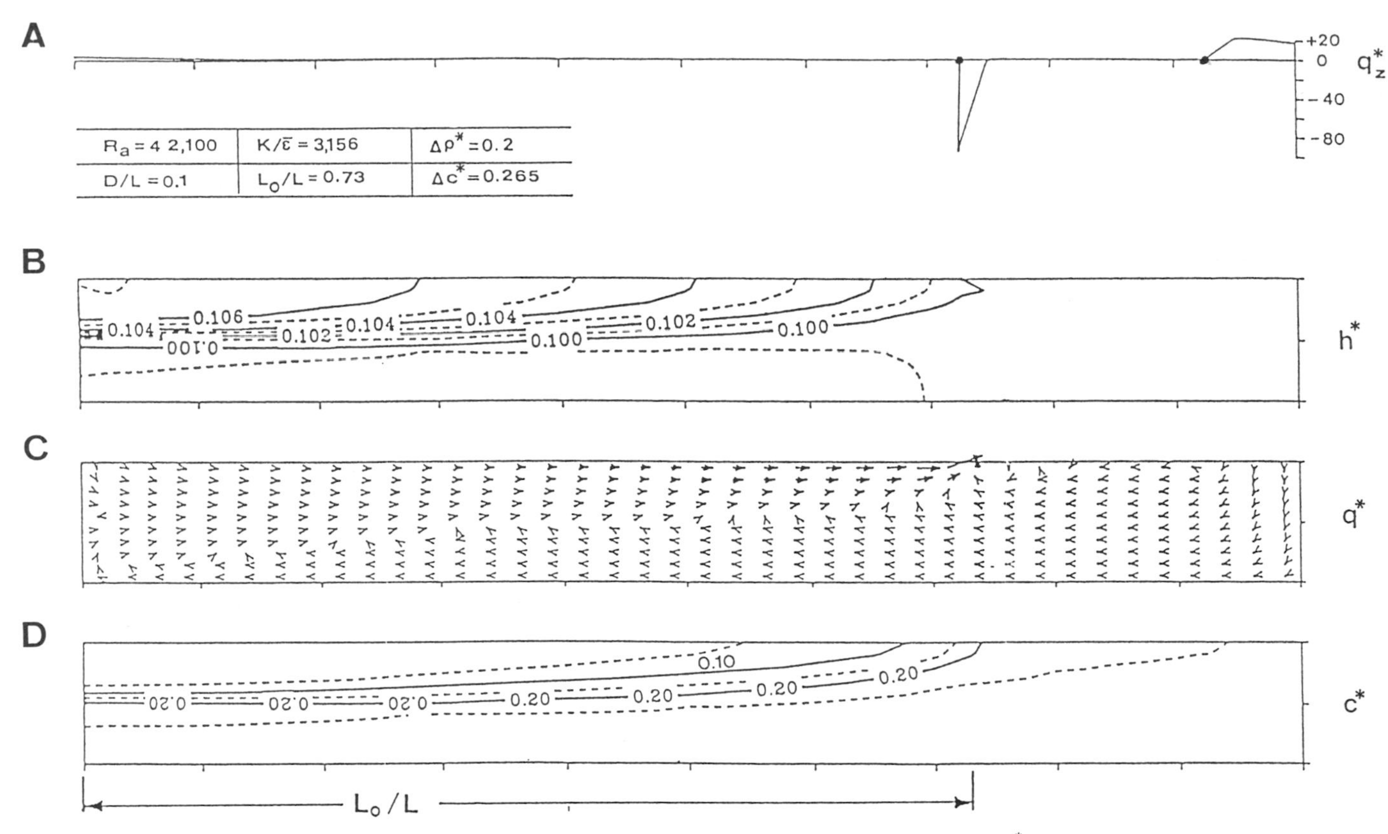

Figure 6. Computer model steady state distribution of: A, recharge-discharge (q_z^*) across the upper boundary; B, the dimensionless hydraulic head ($h^* = p^* + z^*$); C, the normalized velocity field ($q^*(x,z)$); and D, the concentration contours (c^*) that evolve when the basin conductance (K/ε) is large and the density contrast ($\Delta\rho^*$) is used to illustrate a brine that is saturated (from Duffy and Al-Hassan, 1988). The basin-center is to the right. Notice that the velocity field for the basin-center brines is directed down and away from the center of the basin and then upwells and mixes with the less saline margin water.

dition, this type of circulation pattern may also explain the apparent variability of the top of the water table in the playa center and the areas of "self-rising ground" versus flat brine-soaked areas on the saline mud-flat surface.

Barnes et al. (1990) also developed a model in which convection was a main driving component for brine movement. They used a theoretical model combined with field data from four sites in Australia to demonstrate that placing fresher irrigation waste water in discharge lakes on top of natural brines produced by evaporation in the lake may cause a change in the regional groundwater flow paths.

In the natural case, the brine comes to the surface and may advect downward due to density contrasts, but is maintained essentially beneath the lake area as a depression in the regional water table (Fig.7A). When relatively fresh waste water is applied to the surface, it creates a groundwater mound, the pressure of which forces the underlying brine to migrate laterally away from the playa (Fig. 7B). The mounding effect suppresses the regional flow lines creating a through-flow situation in what was originally a groundwater sink (hydrologically closed basin).

They were able to demonstrate that this type of situation occurs at Noora Lake in South Australia. The implication of this study is that these types of playas are not good for storing irrigation waste water because they change the regional groundwater system around the playa from discharging to through-flow. Thus, the irrigation waste water, rather than remaining isolated from the regional groundwater, will eventually recharge into the aquifer system. This study also demonstrates the delicate balance between groundwater flow and evaporation that must be maintained in order to create hydrologically closed basin in areas with very low topographic relief.

DISCUSSION OF CLASSIFICATION SCHEMES

The main problem with the previously presented geomorphic and hydrologic classification schemes of playas was that they relied to some extent on criteria which could have multiple origins. Snyder's (1962) scheme was very simple, but very useful in its compartments (Table 3). He divided the systems up into four basic variables; hydrologically open or closed (drained and undrained), and topographically open or closed.

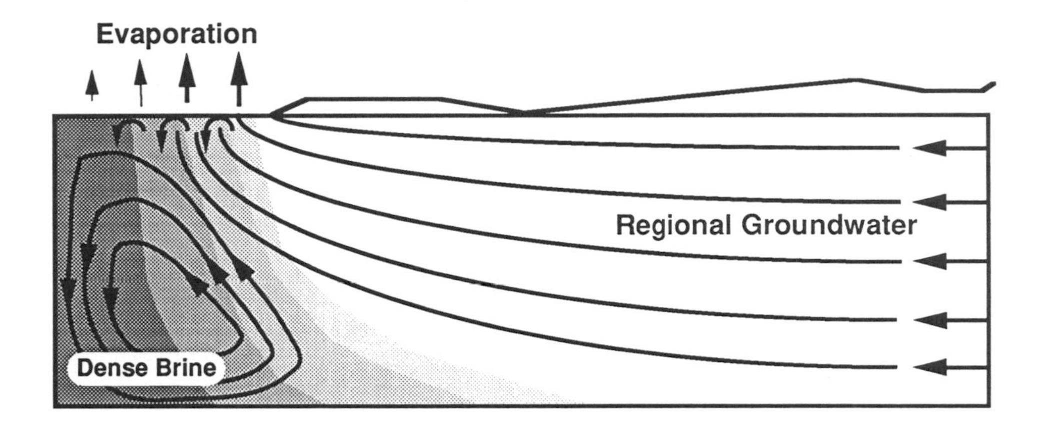

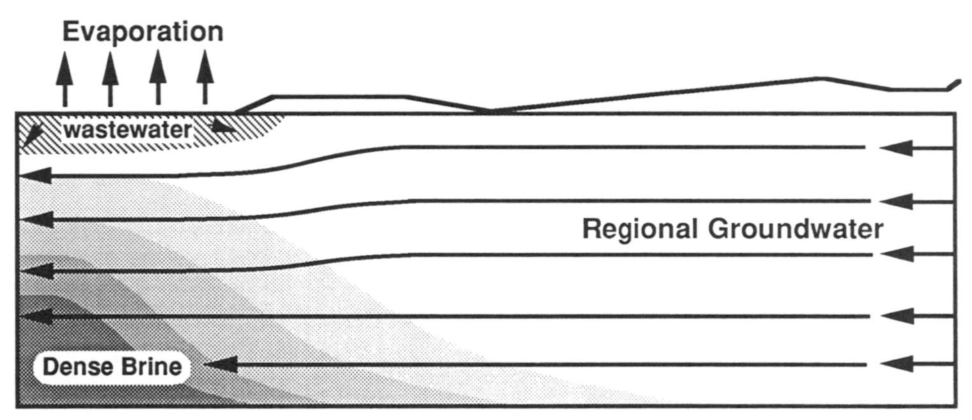

Figure 7. A, Groundwater flow paths in a discharge playa under natural conditions. Notice that the dense brine is located mostly under the discharge zone and that the hydraulic head of the discharge area is lower than regional water table (after Barnes et al., 1990). B, Groundwater flow paths after a freshwater lens has been loaded onto the playa surface. Loading creates an unstable density contrast and imbalance such that mixing may occur. In addition, the mounding of water above the brine raises the water table over the brine creating a downward pressure which causes the groundwater hydrology to change from a discharge lake to a through-flow lake (after Barnes et al., 1990).

He recognized three of the four combinations in the western United States: (1) closed and undrained basins, (2) closed and drained basins, and (3) open and drained basins. He could not find any examples of open undrained basins.

The first compartment, closed and undrained basins, would by anyone's definition contain playas in the lowest part of the depression. These types of basins do not need any further discussion from a hydrological point of view.

Topographically closed but drained basins may contain dry clay pans or seasonally wet pans. However, whether these types of deposits are playas are open to debate. Certainly, it is possible to envision a basin that is topographically closed but has some leakage which is smaller than the net evaporation rate (evaporation minus precipitation). This would lead to an eventual increase in salinity and evaporite deposition. But, because there is some leakage, an equilibrium point would be achieved where the water could not become more saline (Wood and Sanford, 1990; Sanford and Wood, 1991).

Snyder's (1962) third compartment, open and drained basins, is likely to contain the regional discharge complexes mentioned above. These areas would be hydrologically open or drained but could be either topographically open or closed depending on how you defined the drainage basin of the complex. The topographic relief on such basins is so low that it is

TABLE 3. SNYDER'S* CLASSIFICATION SCHEME FOR VALLEYS IN THE GREAT BASIN, WESTERN UNITED STATES

		Topography	
		Valleys are topographically	
		Closed	Open
Hydrologic Regimen	Valleys are hydrologically Drained	1. Alluvium unsaturated near surface but possible water table at depth. 2. Playa dry during dry season. 3. Water escapes through subsurface outlet. 4. Water usually of good quality.	1. Alluvium unsaturated near surface. 2. Water moves out of valley near base of valley fill, or above a partial barrier. 3. Any playa present is usually dry during dry season. 4. Water quality usually good or quality may vary if a partial barrier exists.
	Undrained	1. Alluvium saturated. 2. Playa wet during dry season. 3. Water lost from valley by evapotranspiration. 4. Water quality will range from good away from the playa to poor near the valley center.	Not known in Great Basin.

*Snyder, 1962.

TABLE 4. MABBUTT'S MODIFICATION OF SNYDER'S AND MOTT'S CLASSIFICATION SCHEMES*

Topographically Closed Basin		Topographically Open Basin
No Groundwater Outlet	External Groundwater Outlet	
Moist surface	Moist in wet season	Dry surface
Saline groundwater	Saline or non-saline, depending on climate, topography, and subsurface drainage	Mainly non-saline de-pending on climate and degree of topographic closure
Salt crust or wet mud flats	Silty-clay surface subject to deflation	Clay pan
Mainly large, e.g., Great Salt Lake		Mainly small

→ Increasing importance of surface flow

← Increasing importance of groundwater discharge

*Mabbutt, 1977; Snyder, 1962; Mott, 1965.
Note the importance of groundwater in closed basins.

difficult to determine the degree to which an individual complex is topographically closed or open. Although these low gradient discharge complexes would be the best candidates for a system that was hydrologically open and topographically closed, rain-shadow basins which are not at the terminus of the regional drainage, such as Cadiz Dry Lake or Alkali Dry Lake (California), may also be candidates for this type of system (Rosen, 1991).

One of the components of the hydrological models for playa classification is based on the earlier workers observations of whether the surface of the playa is wet or dry. The modifications of Motts (1965) and Mabbutt (1977) suggest that although playa surfaces may be variable, the degree to which a playa is groundwater dominated or surface-water dominated will determine the characteristics of the hydrology and playa evolution (Table 4). In particular, the ratio of surface water inflow to groundwater inflow is taken to be a determining characteristic. That is, if surface water is dominant then the playa surface will be dry, but if groundwater is dominant then the surface will be wet. In a dry playa, the net aggradation is determined by the flooding ratio. For example, in order to accumulate sediments on a dry playa, the surface must be flooded more often than it is dry, otherwise the sediments would be eroded. Although Mabbutt (1977) and Motts (1965) both admit that most dry playas are now undergoing erosion rather than net deposition, they still argue that dry playas represent a hydrological "type" or class of playas.

It is my contention that dry playas are nothing more than deflation basins which are not at equilibrium with the present hydrologic or climatic conditions. If a basin is truly hydrologically closed, then groundwater must discharge at or near the surface and evaporate. If the basin is hydraulically closed and dominated by surface runoff, then the lake will not be a playa but will be perennial (although possibly saline) and perhaps overflow. An example of this type of basin would be Mono Lake in California. However, as the presence of extensive tufa mounds implies, Mono Lake is also greatly effected by groundwater inflow.

If a basin is topographically closed and the surface is dry, suggesting that groundwater does not discharge in the basin, then the groundwater system must be to a large degree open. In fact, dry playas may actually act as recharge zones to the deep aquifer if a sufficient amount of precipitation falls (Wood and Osterkamp, 1987).

It is impossible for a basin that is hydrologically closed (i.e., it is the terminus of a drainage system) not to be at least moist in the basin center. This does not mean that all basins that are moist are hydraulically closed, but only that those basins that are dry must be either hydraulically open or at some disequilibrium with the present climatic conditions.

In any event, dry playas are only geomorphic features and are unlikely to be preserved in the rock record because, as mentioned above, they are deflation surfaces in the process of eroding (Blackwelder, 1931). Although they are resistant to wind erosion due to their smooth, hard surface (Mabbutt, 1977), they will eventually erode if the water table remains deep. In the rock record, a dry playa would be at best represented by eolian or debris flow sediments, and at worst as a disconformity surface with perhaps a reddened soil beneath it. Therefore, dry basins are classified as recharge playas, which are distinct from discharge types of playas.

The important point of this discussion is that the classification of playas using surface conditions of the playa itself (wet or dry, smooth or pustular) must be tied to a better understanding of the hydrologic regime of the basin. But if this is the case, then perhaps a classification based on the hydraulics of the basin would be more accurate. Unfortunately, details of the basin hydraulics in arid regions are generally not well constrained and are based more on proxy evidence than on actual hydrologic measurements.

Figure 8 summarizes my modifications of the existing hydrological classifications of playas and Table 5 presents a hydrologically based classification of playas. Basins are divided into either hydrologically open or closed systems. However, if the basin is either locally or regionally discharging groundwater it a playa. If water infiltrates to the aquifer through the basin center depression, it is called a recharge playa. Based on Wood and Sanford's (1990) concept of leakage ratios (net outflow over net inflow), discharge playas have no leakage (net outflow = 0), and are generally the terminus of a regional groundwater systems. Through-flow playas have a variable amount of leakage (net outflow is smaller than net inflow),

TABLE 5. A CLASSIFICATION OF PLAYAS BASED ON THE HYDROLOGY OF BASIN*

			Hydrologically Open	Hydrologically Closed	
Capillary Fringe Near Surface	Leakage ratio = 0	Regional groundwater discharge		Discharge playa Evaporites accumulate	Playas
	Leakage ratio <1 and >0	Local groundwater discharge	Through-flow playa Evaporites may accumulate		
Capillary Fringe Below Surface	Leakage ratio = 1	Infiltation to aquifer (recharge)	Recharge playa No evaporites		Recharge Playas

*Definition of leakage ratio is from Wood and Sanford, 1990. Leakage ratio = net outflow over net inflow.

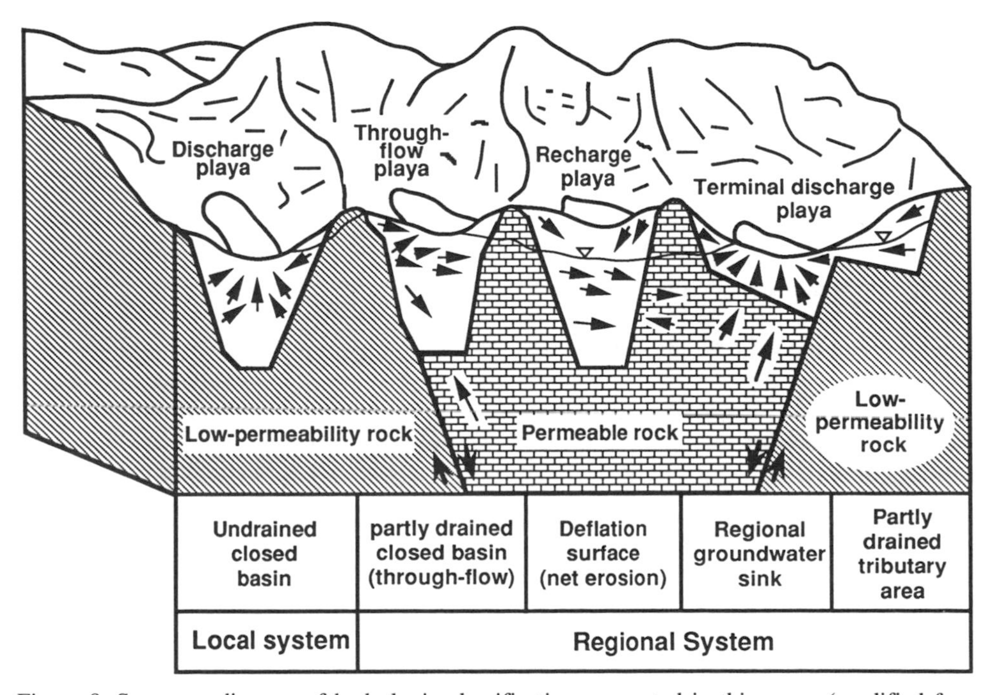

Figure 8. Summary diagram of hydrologic classifications presented in this paper (modified from Eakin et al., 1976). Rather than the term dry playa, the term recharge playa has been substituted.

and in recharge basins, the net outflow is equal to the net inflow (leakage ratio = 1). Although playas are subdivided into through-flow and discharge playas depending on the leakage ratio of the basin, this distinction may not be necessary, and either type of basin may be called simply a playa.

The importance of evaporation is represented by the position of the capillary fringe in the two types of playas. Discharge playas have a capillary fringe near the surface, implying that evaporation is important, and recharge playas have a deep capillary fringe, implying infiltration is dominant.

SUMMARY

The main purpose of this paper has been to establish the importance of understanding the interaction between the geology and sedimentology of playa basins and the dynamics and chemistry of the groundwater entering the basin. The fundamental points I have emphasized are:

(1) Playas and playa-lakes are formed in both hydrologically closed or open basins where evaporation exceeds inflow. They may occur in the zone of local or regional discharge, and may be the terminus of large or small internally drained basins.

(2) In hydrologically closed basins, the lithology of the sediments and bedrock within the playa catchment are important in determining the ultimate chemistry of the brine. However, in hydrologically open (through-flow) discharge complexes, lithology may be subdominant to the composition of the precipitation entering the basin and the leakage ratio of the system.

(3) Saline minerals form in all parts of the basin and recharge area, although significant accumulations of saline minerals will be in the deepest part of the basin. Saline minerals form as (1) surface efflorescent crusts, (2) subsurface intrasedimentary displacive crystals, layers, or nodules, (3) direct precipitates from a standing water body, (4) intra- or extraclasts, and (5) eolian processes. Methods 1 to 3 are directly related to the characteristics of groundwater flow into the basin.

(4) Depressions not connected to the groundwater table are in disequilibrium and will eventually erode down to the level of water-saturated sediments (i.e., the capillary zone) or fill with nonplaya sediments. These depressions do not accumulate evaporite minerals and it is suggested they should be called recharge playas. It is further suggested that the terms *sabkha* and *salina* should be reserved for coastal marine-influenced settings.

(5) Extensive accumulations of subaqueous evaporites form in playas when there is sufficient groundwater to maintain a shallow brine pond for an extended period of time. Extremely arid periods, when there is insufficient groundwater to maintain a brine, will result in displacive intrasediment growth of evaporites and extremely wet periods may be too fresh for a brine to develop. It is only in the intermediate periods when evaporation is high and groundwater input is sufficient that subaqueous evaporites will accumulate in hydrologically closed basins. This has important implications for paleoclimate reconstructions of these types of playa sequences.

(6) The surface morphology and subsequent diagenesis of playa sediments are directly related to the hydrology of the basin. Therefore classifications of playa systems should be based only on the hydrology of the system rather than the geomorphology of the surface. It should be stressed, then, that facies relationships in playa lakes (and to a lesser extent saline lakes) are determined by hydrology, and not necessarily by Walther's Law. Thus, the stratigraphic record of a playa sequence is a reflection of the facies mosaic controlled by a dynamic groundwater system.

ACKNOWLEDGMENTS

I would like to thank Warren Wood and Blas Valero-Garcés for insightful reviews of this manuscript. Their efforts greatly improved the quality and organization of this paper. This paper was written while I was a Research Associate on a National Science Foundation Research Training Grant "Paleorecords of Global Change: Understanding the Dynamics of Ecosystem Response," number DIR 9014277. This paper is contribution No. 455 to the Limnological Research Center, University of Minnesota.

REFERENCES CITED

Arndt, J. L., and Richardson, J. L., 1993, Temporal variations in the salinity of shallow groundwater from the periphery of some North Dakota wetlands (USA): Journal of Hydrology, v. 141, p. 75–105.

Barnes, C. J., Chambers, L. A., Herczeg, A. L., Jacobson, G., Williams, B. G., and Wooding, R. A., 1990, Mixing processes between saline groundwater and evaporation brines in groundwater discharge zones, *in* Proceedings of the International Conference on Groundwater in Large Sedimentary Basins, Perth: Perth, Australian Water Resources Council, Conference Series No. 20, p. 369–378.

Bear, J., 1979, Hydraulics of groundwater: New York, McGraw-Hill Inc., 567 p.

Blackwelder, E., 1931, The lowering of playas by deflation: American Journal of Science, 5th Serial, v. 21, p. 140–144.

Butler, G. P., 1969, Modern evaporite deposition and geochemistry of coexisting brines, the sabkha, Trucial Coast, Arabian Gulf: Journal of Sedimentary Petrology, v. 39, p. 70–89.

Castens-Seidell, B., 1984, The anatomy of a modern marine siliciclastic sabkha in a rift valley setting: Northwest Gulf of California tidal flats, Baja California, Mexico [Ph.D. thesis]: Johns Hopkins University, 386 p.

Chivas, A. R., Andrew, A. S., Lyons, W. B., Bird, M. I., and Donnelly, T. H., 1991, Isotopic constraints on the origin of salts in Australian playas. 1. Sulphur: Palaeogeography, Palaeoclimatology, Palaeoecology, v. 84, p. 309–332.

Christian, C. S., Jennings, J. N., and Twidale, C. R., 1957, Geomorphology, *in* Dickson, B. T., Arid zone Research-IX, Guide Book to research data for arid zone development, UNESCO; Frankfurt, J. Weisbecker, p. 51–65.

Clarke, F. W., 1924, The data of geochemistry: U.S. Geological Survey Bulletin, 770, 841 p.

Commander, D. P., Kern, A. M., and Smith, R. A., 1991, Hydrology of the Tertiary palaeochannels in the Kalgoorlie Region (Roe Palaeodrainage): Geological Survey of Western Australia Hydrological Record 1991/10, 56 p.

Curtis, R., Evans, G., Kinsman, D.J.J., and Shearman, D. J., 1963, Association of dolomite and anhydrite in the recent sediments of the Persian Gulf: Nature, v. 197, p. 6779–6800.

Donovan, J. J., and Rose, A. W., 1992, The chemical jump: Hydrologic control of brine reaction path within calcareous aquifers and alkaline lakes, semi-arid northern Great Plains, USA, *in* Kharaka, Y. K., and Maest, A. S., eds., Water Rock Interaction, Vol. 1, Low Temperature Environments, Proceedings of the 7th International Symposium on Water Rock interaction, Park City, Utah: Rotterdam, Balkema, p. 639–642.

Duffy, C. J., and Al-Hassan, S., 1988, Groundwater circulation in a closed desert basin: Topographic scaling and climatic forcing: Water Resources Research, v. 24, p. 1675–1688.

Eakin, T. E., Price, D., and Harrill, J. R., 1976, Summary appraisals of the nation's ground-water resources—Great Basin region: U.S. Geological Survey Professional Paper 813-G, 37 p.

Eugster, H. P., and Hardie, L. A., 1975, Sedimentation in an ancient playa-lake complex: the Wilkins Peak Member of the Green River Formation of Wyoming: Geological Society of America Bulletin, v. 86, p. 319– 334.

Eugster, H. P., and Hardie, L. A., 1978, Saline lakes, in Lerman, A., ed., Lakes: Chemistry, geology, physics: New York, Springer-Verlag, p. 237–293.

Evans, G., Schmidt, V., Bush, P., and Nelson, H., 1969, Stratigraphy and geologic history of the sabkha, Abu Dhabi, Persian Gulf: Sedimentology, v. 12, p. 145–159.

Ferguson, J., Jacobson, G., Evans, W. R., White, I. Wooding, R. A., Barnes, C. J., and Tyler, S., 1992, Advection and diffusion of groundwater brines in modern and ancient salt lakes, Nulla groundwater discharge complex, Murray Basin, southeast Australia, in Kharaka, Y. K.,and Maest, A. S., eds., Water Rock Interaction, Vol. 1, Low Temperature Environments, Proceedings of the 7th International Symposium on Water Rock interaction, Park City, Utah: Rotterdam, Balkema, p. 643–647.

Foshag, W. F., 1926, Saline lakes of the Mojave desert: Economic Geology, v. 21, p. 56–64.

Gilbert, G. K, 1875, The Glacial Epoch, Chapter III, *in* v. III (geology) (geology), Report on geographical and geological surveys west of the 100th meridian: Washington, D.C., Government Printing Office, p. 86–104.

Handford, C. R., 1981, A process-sedimentary framework for characterizing recent and ancient sabkhas: Sedimentary Geology, v. 30, p. 255–265.

Handford, C. R., 1982a, Sedimentology and evaporite genesis in a Holocene continental-sabkha playa basin—Bristol Dry Lake, California: Sedimentology, v. 29, p. 239–253.

Handford, C. R., 1982b, Terrigenous clastic and evaporite sedimentation in a Recent continental-sabkha playa basin, Bristol Dry Lake, California, *in* Handford, C. R., Loucks, R. G., and Davies, G. R., eds., Depositional and Diagenetic Spectra of Evaporites—A core workshop: Calgary, Canada, Society of Economic Paleontologists and Mineralogists Core Workshop No. 3, p. 65–74.

Hardie, L. A., 1968, The origin of the Recent non-marine evaporite deposit of Saline Valley, Inyo County, California: Geochimica et Cosmochimica et Cosmochimica Acta, v. 32, p. 1279–1301.

Hardie, L. A., 1984, Evaporites: Marine or non-marine?: American Journal of Science, v. 284, p. 193–240.

Hardie, L. A., 1990, The role of rifting and hydrothermal $CaCl_2$ brines in the origin of potash evaporites: An hypothesis: American Journal of Science, v. 290, p. 43–106.

Hardie, L. A., and Eugster, H. P., 1970, The evolution of closed-basin brines: Special Paper Mineralogical Society of America, v. 3, p. 273–290.

Hardie, L. A., Smoot, J. P., and Eugster, H. P., 1978, Saline lakes and their deposits: A sedimentological approach, in Matter, A., and Tucker, M. E., eds., Modern and Ancient Lake Sediments: International Association of Sedimentologists Special Publication, v. 2, p. 7–41.

Herczeg, A. L., and Lyons, W. B., 1991, A chemical model for the evolution of Australian sodium chloride lake brines: Palaeogeography, Palaeoclimatology, Palaeoecology, v. 84, p. 43–53.

Holser, W. T., 1979, Mineralogy of evaporites, in Burns, R. G., ed., Marine Minerals: Reviews in Mineralogy, v. 6, Mineralogical Society of America, p. 211–294.

Jacobson, G., and Jankowski, J., 1989, Groundwater-discharge processes at a central Australian playa: Journal of Hydrology, v. 105, p. 275–295.

Jacobson, G., Jankowski, J. and Chen, X. Y., 1991, Solute budget for an arid–zone groundwater system, Lake Amadeus, central Australia: Australian Journal of Earth Sciences, v. 38, p. 1–14.

Jankowski, J., and Jacobson, G., 1989, Hydrochemical evolution of the regional groundwaters to playa brines in central Australia: Journal of Hydrology, v. 108, p. 123–174.

Jones, B. F., 1966, Geochemical evolution of closed basin waters in the western Great basin, *in* Rau, J. L., ed., Second Symposium on Salt: Northern Ohio Geological Society, v. 1, p. 181–200.

Jones, B. F., and Van Denburgh, A. S., 1966, Geochemical influences on the chemical character of closed lakes: International Association of Scientific Hydrology Symposium de Garda, Publication 70, p. 435–446.

Jones, B. F., Eugster, H. P., and Rettig, S. L., 1977, Hydrochemistry of the Lake Magadi Basin, Kenya: Geochimica et Cosmochimica Acta, v. 41, p. 53–72.

Kendall, A. C., 1988, Aspects of evaporite basin stratigraphy, *in* Schreiber, B. C., ed., Evaporites and hydrocarbons: New York, Columbia University Press, p. 11–65.

Kinsman, D.J.J., 1966, Gypsum and anhydrite of recent age, Trucial Coast, Persian Gulf, in Rau, J. L., ed., Second Symposium on Salt: Northern Ohio Geological Society, v. 1, p. 302–306.

Langbein, W. B., 1961, Salinity and hydrology of closed lakes: U.S. Geological Survey Professional Paper 412, p. 20.

Lowenstein, T. K., and Hardie, L. A., 1985, Criteria for the recognition of salt–pan evaporites: Sedimentology, v. 32, p. 627–644.

Lowenstein, T. K., Spencer, R. J. and Pengxi, Z., 1989, Origin of ancient potash evaporites: Clues from the Modern nonmarine Qaidam Basin of western China: Science, v. 245, p. 1090–1092.

Mabbutt, J. A., 1977, Desert Landforms, An introduction to systematic geomorphology, Volume 2: Cambridge, The MIT Press Edition, 340 p.

Macumber, P. G., 1991, Interaction between ground water and surface systems in northern Victoria: Melbourne, Department of Conservation and Environment-Victoria, 345 p.

McArthur, J. M., Turner, J. V., Lyons, W. B., and Thirwall, M. F., 1989, Salt sources and water-rock interaction on the Yilgarn Block, Australia: Isotopic and major element tracers: Applied Geochemistry, v. 4, p. 79–92.

Meinzer, O. E., 1922, Map of the Pleistocene lakes of the Basin-and-Range Province and its significance: Geological Society of America Bulletin, v. 33, p. 541–552.

Meinzer, O. E., 1927, Plants as indicators of groundwater: U.S. Geological Survey Water-Supply Paper 577, 95 p.

Mifflin, M. D., 1968, Delineation of ground-water flow systems in Nevada: University of Nevada, Desert Research Institute, Technical Report Series H-W, v. 4, 54 p.

Motts, W. S., 1965, Hydrologic types of playas and closed valleys and some relations of hydrology to playa geology, *in* Neal, J. T., ed., Geology, Mineralogy, and Hydrology of U.S. Playas, Airforce Research Laboratory: Bedford, Massachusetts, Environmental Research Papers 96, p. 73–105.

Neal, J. T., 1975, Introduction, *in* Neal, J. T., ed., Playas and Dried Lakes: Stroudsburg, Dowden, Hutchinson, and Ross, Inc., Benchmark Papers in Geology, p. 1–5.

Osterkamp, W. R., and Wood, W. W., 1987, Playa-lake basins in the Southern High Plains of Texas and New Mexico: Part I. Hydrologic, geomorphic, and geologic evidence for their development: Geological Society of America Bulletin, v. 99, p. 215–223.

Reeves, C. C., Jr., 1968, Introduction to paleolimnology: Amsterdam, Elsevier, Developments in Sedimentology 11, 228 p.

Reeves, C. C., Jr., 1978, Economic significance of playa lake deposits, *in* Matter, A., and Tucker, M. E., eds., Modern and Ancient Lake Sediments: Oxford, Special Publication No. 2, International Association of Sedimentologists, Blackwell Scientific Publications, p. 279–290.

Rosen, M. R., 1989, Sedimentologic, geochemical, and hydrologic evolution of an intracontinental, closed-basin playa (Bristol Dry Lake, California): A model for playa development and its implications for paleoclimate [Ph.D. thesis]: Austin, University of Texas-Austin, 266 p.

Rosen, M. R., 1991, Sedimentologic and geochemical constraints on the hydrologic evolution of Bristol Dry Lake, California, U.S.A.: Palaeogeography, Palaeoclimatology, Palaeoecology, v. 84, p. 229–257.

Rosen, M. R., and Warren, J. K., 1990, The origin of groundwater-seepage gypsum from Bristol Dry Lake, California, U.S.A.: Sedimentology, v. 37, p. 983–996.

Russell, I. C., 1883, Playas and playa-lakes: The Popular Science Monthly, v. 22, p. 380–383.

Russell, I. C., 1885, Playa-lakes and playas: U.S. Geological Survey Monograph, no. 11, p. 81–86.

Sanford, W. E., and Wood, W. W., 1991, Brine evolution and mineral deposition in hydrologically open evaporite basins: American Journal of Science, v. 291, p. 687–710.

Schmid, R. M., 1988, Lake Torrens halite accumulation (South Australia): Zeitschrift der Deutschen Geologischen Gesellschaft, v. 139, p. 289–296.

Smith, G. I., Friedman, I., and McLaughlin, R. J., 1987, Studies of Quaternary saline lakes—III. Mineral, chemical, and isotopic evidence of salt solution and crystallization processes in Owens Lake, California, 1969–1971: Geochimica et Cosmochimica Acta, v. 51, p. 811–828.

Smoot, J. P., 1983, Depositional subenvironments in an arid closed basin; Wilkins Peak Member of the Green River Formation (Eocene), Wyoming, USA: Sedimentology, v. 30, p. 801–827.

Smoot, J. P., 1991, Sedimentary facies and depositional environments of early Mesozoic Newark Supergroup basins, eastern North America: Palaeogeography, Palaeoclimatology, Palaeoecology, v. 84, p. 369–423.

Smoot, J. P., and Lowenstein, T. K., 1991, Depositional environments of nonmarine evaporites, *in* Melvin, J. L., ed., Evaporites, Petroleum and Mineral Resources: Amsterdam, Elsevier, Developments in Sedimentology 50, p. 189–347.

Snyder, C. T., 1962, A hydrological classification of valleys in the Great Basin, western U.S.A.: International Association of Scientific Hydrology, v. 7, p. 53–59.

Teller, J. T., and Last, W. M., 1990, Paleohydrological indicators in playas and salt lakes, with examples from Canada, Australia, and Africa: Palaeogeography, Palaeoclimatology, Palaeoecology, v. 76, p. 215–240.

Thompson, D. G., 1929, The Mojave Desert region, California: U.S. Geological Survey Water-Supply Paper 578, 759 p.

Townley, L. R., and Davidson, M. R., 1988, Definition of a capture zone for shallow water table lakes: Journal of Hydrology, v. 104, p. 53–76.

Townley, L. R., and Turner, J. V., 1992, Groundwater flow near shallow lakes: New insights and implications for management: Australian Water and Wastewater Association Water Journal, v. 19, p. 40–41.

Townley, L. R., Barr, A. D., and Nield, S. P., 1992, FlowThru: An interactive program for calculating groundwater flow regimes near shallow surface water bodies: version 1.0: CSIRO-Australia, Division of Water Resources, Technical Memorandum 92/1, 27 p.

Turner, J. V., Townley, L. R., Rosen, M. R., and Sklash, M. K., 1992, Coupling the spatial distribution of solute concentration and stable isotope enrichments to hydrologic processes, *in* Kharaka, Y. K., and Maest, A. S., eds., Water Rock Interaction Vol. 1, Low Temperature Environments, Proceedings of the 7th International Symposium on Water Rock interaction, Park City, Utah: Rotterdam, Balkema, p. 217–221.

Warren, J. K., 1982a, The hydrologic setting, occurrence and significance of gypsum in late Quaternary salt lakes in South Australia: Sedimentology, v. 29, p. 609–637.

Warren, J. K., 1982b, The hydrological significance of Holocene tepees, stromatolites and boxwork limestones in coastal salinas in South Australia: Journal of Sedimentary Petrology, v. 52, p. 1171–1201.

Warren, J. K., and Kendall, G.ST.C., 1985, Comparison of sequences formed in marine sabkha (subaerial) and saline (subaqueous) settings—Modern and ancient: American Association of Petroleum Geologists Bulletin, v. 69, p. 1013–1023.

Winter, T. C., 1976, Numerical simulation analysis of lakes and groundwater: U.S. Geological Survey Professional Paper 1001, 45 p.

Winter, T. C., 1978, Numerical simulation of steady state three-dimensional groundwater flow near lakes: Water Resources Research, v. 14, p. 245–254.

Winter, T. C., and Carr, M. R., 1980, Hydrologic setting of wetlands in the Cottonwood Lake area, Stutsman County, North Dakota: U.S. Geological Survey Water Resource Investigation Report 80-99, 42 p.

Wood, W. W., and Osterkamp, W. R., 1987, Playa-lake basins in the Southern High Plains of Texas and New Mexico: Part II. A hydrologic model and mass-balance arguments for their development: Geological Society of America Bulletin, v. 99, p. 224–230.

Wood, W. W., and Sanford, W. E., 1990, Ground-water control of evaporite deposition: Economic Geology, v. 85, p. 1226–1235.

Manuscript Accepted by the Society July 2, 1993

Printed in U.S.A.

Geological Society of America
Special Paper 289
1994

Major-element and stable-isotope geochemistry of fluid inclusions in halite, Qaidam Basin, western China: Implications for late Pleistocene/Holocene brine evolution and paleoclimates

Tim K. Lowenstein
Department of Geological Sciences, State University of New York at Binghamton, Box 6000, Binghamton, New York 13902-6000
Ronald J. Spencer and Yang Wenbo
Department of Geology and Geophysics, The University of Calgary, Calgary, Alberta, T2N 1N4 Canada
Enrique Casas
Department of Geological Sciences, State University of New York at Binghamton, Box 6000, Binghamton, New York 13902-6000
Zhang Pengxi, Zhang Baozhen, and Fan Haibo
Institute of Salt Lakes, Academia Sinica, Xining, Qinghai Province, People's Republic of China
H. Roy Krouse
Department of Physics, The University of Calgary, Calgary, Alberta, T2N 1N4 Canada

ABSTRACT

Salt cores from the nonmarine Qaidam Basin, western China, are used to interpret late Pleistocene/Holocene paleoclimate and brine evolution based on major elements and stable isotopes (hydrogen and oxygen) in fluid inclusions in halite. Layered primary halite with chevron and cumulate textures diagnostic of crystallization at the brine bottom and at the air-brine interface occurs to depths of 43 m (54 kyr B.P. based on U-series dating methods). Fluid inclusions in this halite contain the lake brines from which the halite crystallized. Variations in the major-element chemistry and activity of H_2O in fluid inclusions with depth/time reflect changes in the ratio of inflow to evaporation and document relative basin aridity. Stable isotopes of hydrogen and oxygen in fluid inclusion waters are used to interpret paleoclimate in terms of surface temperatures, although atmospheric circulation patterns, sources of moisture, and other factors may also influence isotopic compositions. Relative CO_2/N_2 in fluid inclusions from cumulate halite that contain trapped air preserve a record of atmospheric chemistry.

Fluid inclusions in halite crystals 20 to 13 kyr B.P. in age have exceptionally low $\delta^{18}O$, δD, and CO_2, which coincides with the marine oxygen isotope record and the polar ice core record of global cooling during the last glacial period. At this time, halite precipitated from concentrated saline lake brines in the cool, but hyperarid, Qaidam Basin. Fluid inclusions in salt cores suggest that regional aridity may have limited glaciation on the Qinghai-Tibet Plateau during the last glacial, compared to North America and other areas.

Lowenstein, T. K., Spencer, R. J., Yang, W., Casas, E., Zhang, P., Zhang, B., Fan, H., and Krouse, H. R., 1994, Major-element and stable-isotope geochemistry of fluid inclusions in halite, Qaidam Basin, western China: Implications for late Pleistocene/Holocene brine evolution and paleoclimates, *in* Rosen, M. R., ed., Paleoclimate and Basin Evolution of Playa Systems: Boulder, Colorado, Geological Society of America Special Paper 289.

Fluid inclusions in halite from 13 kyr B.P. to the present in age have increased $\delta^{18}O$, δD, salinity, and CO_2, indicating probable warming during the glacial-interglacial transition and the Holocene; warm, arid conditions have persisted to the present. The modern Qaidam Basin, which contains the world's largest modern deposits of potash evaporites, represents the driest climate of the 50-kyr-B.P. record.

INTRODUCTION

Evaporites and associated closed-basin sediments, because of the special climatic and hydrologic conditions necessary for their formation and accumulation, hold great potential for understanding processes ranging from mechanisms of brine evolution (Eugster and Hardie, 1978; Hardie et al., 1978; Hardie, 1984; among many others) to detailed reconstruction of paleoclimates. The continental climate record from Plio-Pleistocene evaporite deposits at Searles Lake, California, for example, has been compared with the marine oxygen isotope record by Smith (1984), Bischoff et al. (1985), and Jannik et al. (1991). One aspect that has been relatively ignored in such studies is fluid inclusions. Saline minerals commonly trap fluid inclusions during their precipitation, either subaqueously in the sedimentary environment, or in the pore fluids of the subsurface diagenetic environment. Certain fluid inclusions in crystals of halite therefore carry information on the chemical and isotopic composition of the surface brines from which the halite precipitated, whereas others contain the pore fluid brines present when diagenetic halite grew. The two types of fluid inclusions can be distinguished petrographically (Lowenstein and Hardie, 1985; Casas and Lowenstein, 1989). Borehole cores through halite-bearing evaporites therefore preserve a record of the chemistry and evolution of surface brines and subsurface diagenetic brines.

This paper summarizes the results of petrographic and fluid inclusion studies from two 50-m-long borehole cores through halite and potash salt bearing evaporites from the Qaidam Basin of western China. The major-element and stable isotope (H and O) chemistry of fluid inclusions in crystals of halite, in conjunction with petrographic analysis, are used to interpret brine history and paleoclimate of the Qaidam Basin over the past 50 kyr B.P. Relative CO_2/N_2 in fluid inclusions from halites that contain trapped air preserve a record of atmospheric chemistry. The paleoclimate record from the Qaidam Basin is compared with ice-core data obtained from Antarctica and from the Dunde Ice Cap in the Qilian Mountains, which border the Qaidam Basin to the north (Thompson et al., 1989). Documenting the temporal evolution of brines in the Qaidam Basin has helped clarify the mechanisms of formation of potash evaporite deposits (Casas et al., 1992). Information on the late Pleistocene and Holocene climate of western China may help explain the scarcity of glacial moraines there compared with Europe and North America (Zheng, 1989; Burbank and Kang, 1991). Furthermore, because of its location on the northern margin of the Qinghai-Tibet Plateau, the Qaidam Basin has broader significance in understanding the influence of large plateau regions on global climate (Ruddiman and Kutzbach, 1989; Manabe and Broccoli, 1990).

The Qaidam Basin (36 to 39°N lat.) is a nonmarine closed-basin in western China on the northern margin of the Qinghai-Tibet Plateau (Fig. 1). The basin center has an average elevation of 2,800 m above sea level and the surrounding Kunlun and Qilian Mountains rise to heights of over 5,000 m above sea level. The Qaidam Basin is notable for its aridity, size (>120,000 km^2), hydrocarbon resources (Wang and Coward, 1990), and because it contains the world's largest accumulations of modern potash-bearing (carnallite:$KCl \cdot MgCl_2 \cdot 6H_2O$) evaporites. Mesozoic and Cenozoic continental sediments occur to depths of at least 14 km, including 1,000 m of Quaternary muds that overlie up to 6,000 m of Pliocene evaporite-bearing nonmarine sediments (Chen and Bowler, 1986; Wang and Coward, 1990). The latest phase of deposition (late Pleistocene and Holocene) has involved the formation of vast evaporite deposits, including potash salts, in the central portions of the basin. The largest modern area underlain by evaporites is the 6,000 km^2 Qarhan salt plain (Fig. 1; Chen and Bowler, 1985, 1986; Zhang, 1987). The surface of Qarhan is predominantly dry and consists of a solid crust of halite, with saline groundwater at depths of 0 to 130 cm. Qarhan contains ten shallow (less than 1 m deep) perennial and ephemeral saline lakes, the largest of which is the 200-km^2 Dabusun Lake. Dabusun Lake brines are Na-Mg-Cl-rich and saturated with halite; halite and carnallite are presently accumulating on the restricted margins of the lake (Casas et al., 1992).

Inflow waters to Qarhan and their evolution into concentrated brines capable of precipitating potash salts have been studied by Zhang (1987), Lowenstein et al. (1989), and Spencer et al. (1990b). Inflow consists of perennial river waters (e.g., the Golmud River) and spring waters, which reach the surface at the northern margin of Qarhan along a linear karsted fault zone (Fig. 1). The saline groundwaters and lake waters at Qarhan are produced by mixing varying amounts of river and spring inflow, followed by evaporative concentration (Lowenstein et al., 1989). The river waters entering Qarhan are enriched in Na-HCO_3 and similar in chemical composition to average world river waters except Qarhan rivers have relatively more Na and Cl and less Ca. The river waters are meteoric in origin and derive their solutes from low-temperature weathering reactions involving bedrock and rainwater. Spring inflow waters contain Na as the dominant cation, with lesser Ca, Mg, and K; Cl is the only significant anion; the spring waters con-

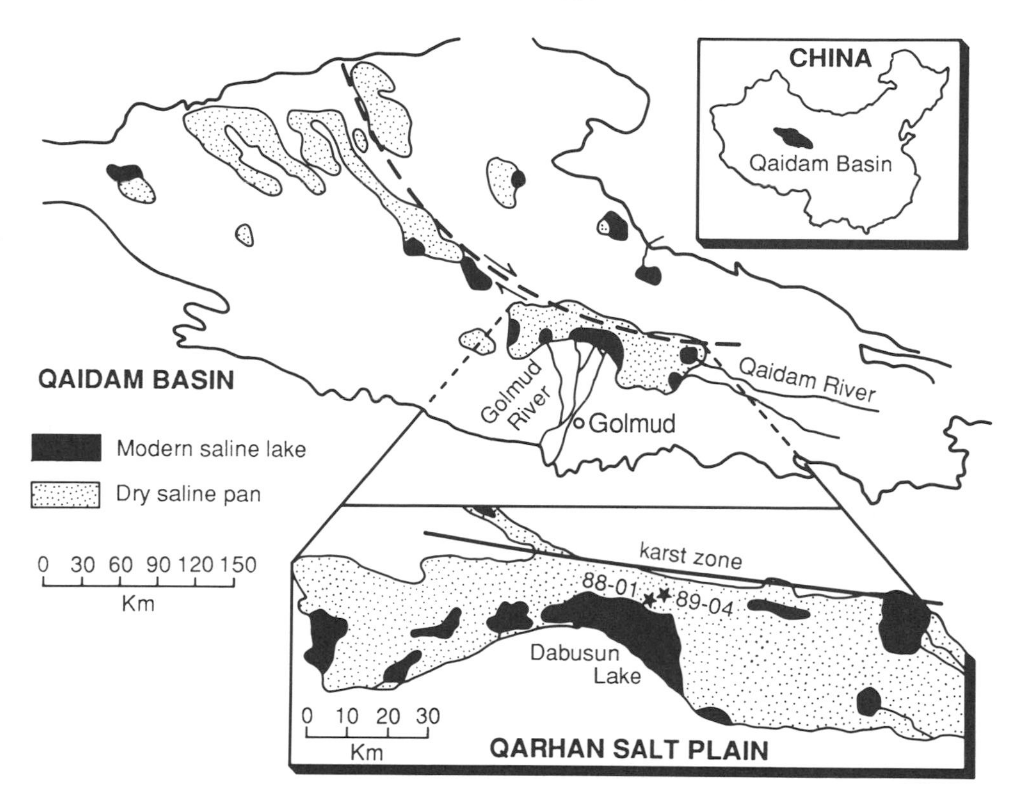

Figure 1. The Qaidam Basin, with locations of cores 88-01 and 89-04. Dashed line through northern margin of Qarhan and extending northwest across the Qaidam Basin is a fault zone along which springs are found. Marginal areas are mostly occupied by alluvial fans and dunes. Modified from Zhang (1987).

tain virtually no SO_4 or HCO_3. The spring inflow may be considered a CaCl brine because it contains more Ca on an equivalent basis than $SO_4 + HCO_3 + CO_3$. Qarhan spring inflow water is similar chemically to many saline formation waters and hydrothermal brines, and has been interpreted to originate in the subsurface by circulation and interaction of hot groundwater with sediments or rocks (Lowenstein et al., 1989).

METHODS AND RESULTS

Late Pleistocene and Holocene evaporites occur to depths of up to 70 m at Qarhan. These deposits have been examined in numerous borehole cores drilled for exploration of potash-salt resources (Chen and Bowler, 1985, 1986; Bowler et al., 1986; Wu et al., 1986; Zhang, 1987; Casas et al., 1992). Two continuous borehole cores drilled in 1988 and 1989 constitute the material of this study (Fig. 2). The cores were taken on the dry salt crust of Qarhan just north of Dabusun Lake (Fig. 1). They are composed predominantly of interlayered halite and mud, like all other cores from this area (Chen and Bowler, 1985, 1986; Bowler et al., 1986; Wu et al., 1986; Lowenstein et al., 1989; Casas et al., 1992). U-series age dating of halites from cores 88-01 and 89-04 has been done at the Institute of Salt Lakes (Fig. 2); the basal halites at 43 m have an age of 54 kyr B.P. (Zhang et al., 1990; Zhang Pengxi, written communication, 1990). Zhang et al. (1990) derived an equation for age versus depth (1.01 kyr/m + 6.16) from core 88-01. The U-series ages are considerably older than those derived from carbon-14 dating of laminated muds from nearby borehole cores (e.g., 19.8 m = 16.4 kyr B.P.; 47.0 m = 26.4 kyr B.P., see Bowler et al., 1986). However, Bowler et al., (1986, p. 255) warn that the carbon-14 dates may not be reliable because "the possibility of contamination by modern carbon remains a constant hazard when dealing with samples so low in organic content." We have relied on U-series dating of halite from cores 88-01 and 89-04. Ages referred to in the text are derived from the equation of Zhang et al (1990).

Late Pleistocene and Holocene siliciclastic muds and halites from Qarhan were examined during core drilling; thin sections were prepared of every interval with layered halite (Fig. 2). Petrographic textures formed the basis for distinguishing primary depositional halite from diagenetic halite which, in turn, aided in the selection of samples for fluid inclusion analysis.

Layered primary halite consists of millimeter- to centimeter-sized crystals with dense fluid inclusion banding (Fig. 3A). Primary halite occurs as chevrons with a vertical growth fabric and textures diagnostic of precipitation at the bottom of a brine lake, and as cumulates consisting of a framework of small crystal plates, linked rafts, and small cubes, all of which grew at the air-brine interface and settled to the bottom of a brine body (Lowenstein and Hardie, 1985). Layers of chevron halite and cumulate halite commonly contain a later genera-

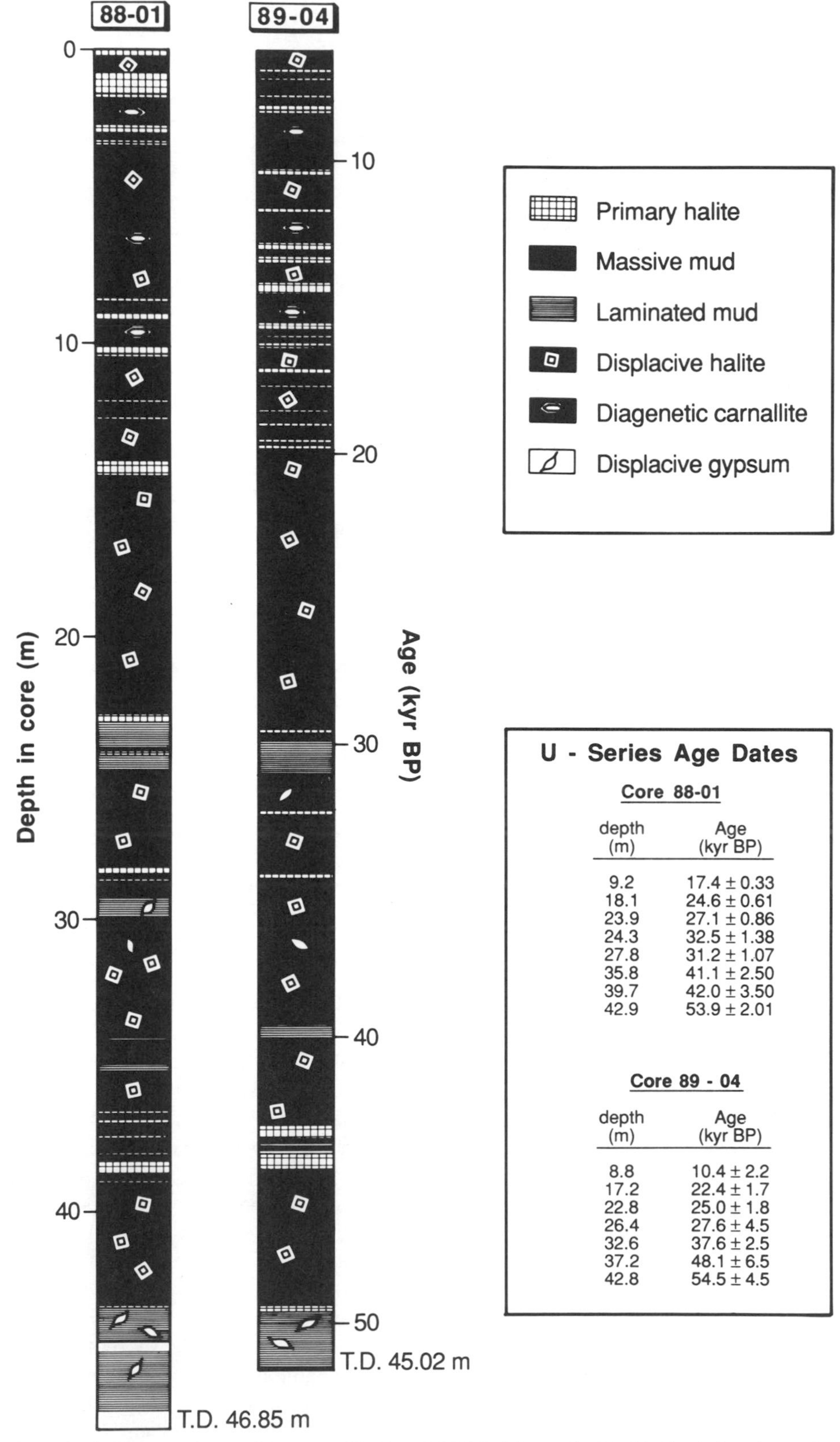

Figure 2. Sediments observed in cores 88-01 and 89-04. Primary bedded halite is interlayered with mud. Diagenetic halite as displacive crystals and cement is abundant at all depths. Diagenetic carnallite only occurs in the top 13 m. Age dates were obtained by uranium-series methods at the Institute of Salt Lakes, Xining, China, and are plotted using the equation age = 1.01 kyr/m + 6.16 from Zhang et al. (1990).

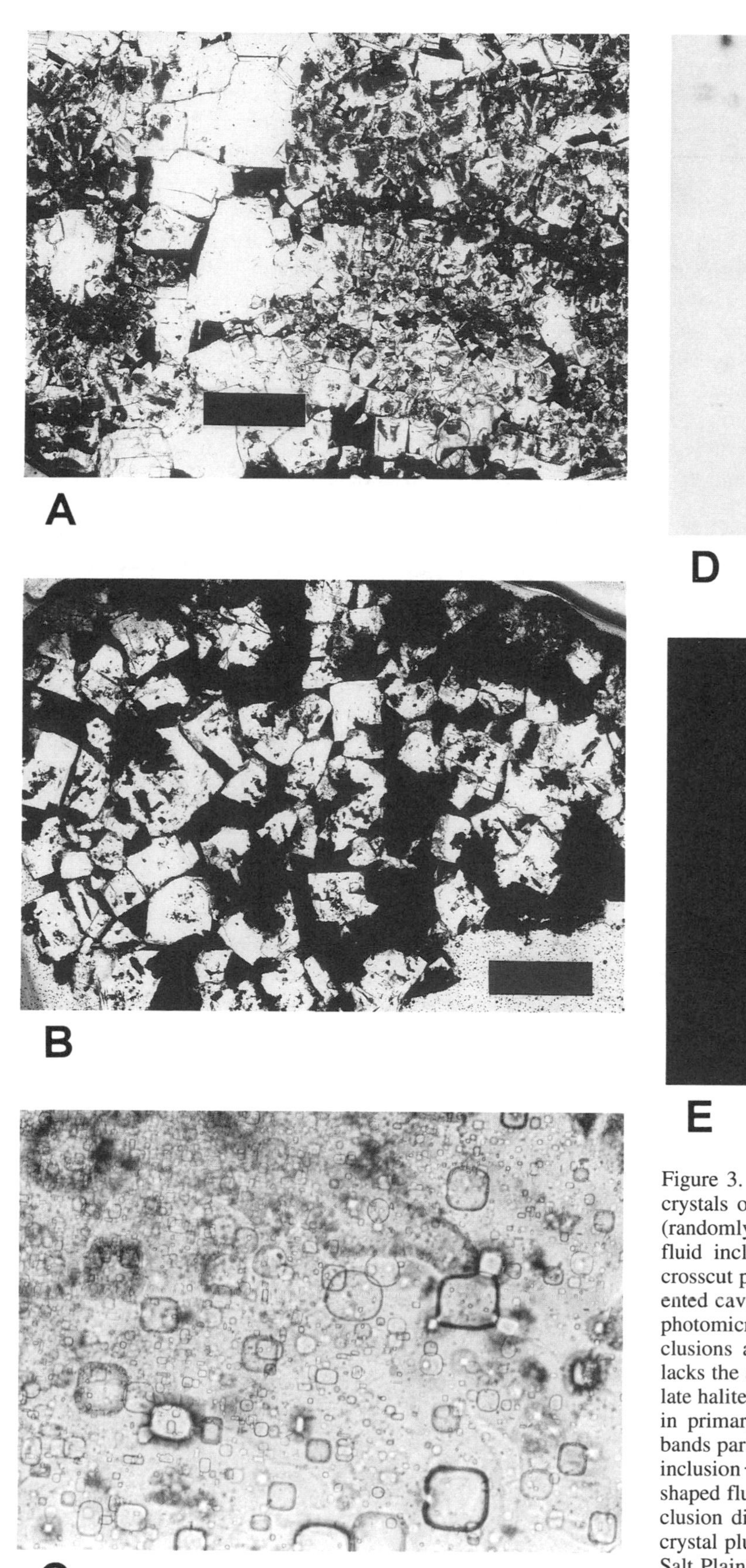

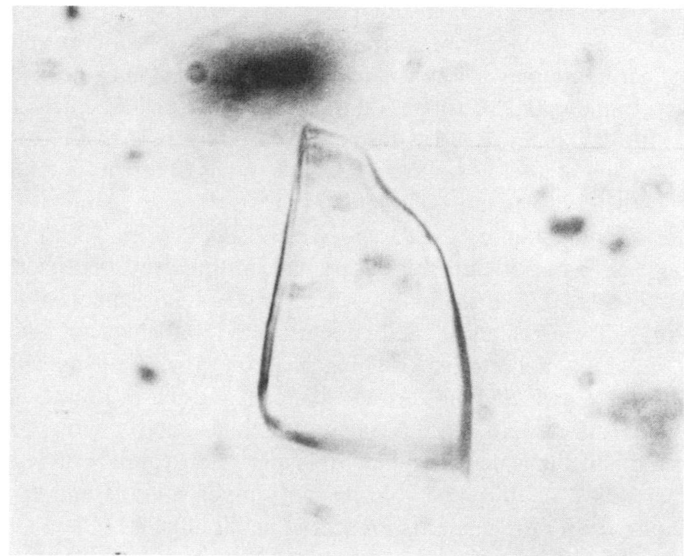

Figure 3. A, Thin-section photomicrograph of primary halite. Small crystals of chevron halite (vertically elongated) and cumulate halite (randomly oriented cubes) have dark cores due to the abundance of fluid inclusions. Coarse crystals of clear diagenetic halite, which crosscut primary halite, are diagenetic cements filling a vertically oriented cavity. Core 88-01, 38.3 m; scale bar is 1 cm. B, Thin-section photomicrograph of displacive halite cubes grown in mud. Mud inclusions are incorporated in the halite crystals. This type of halite lacks the abundant fluid inclusions typical of primary chevron/cumulate halite. Core 88-01, 24.2 m; scale bar is 1 cm. C, Fluid inclusions in primary chevron halite are characteristically densely packed in bands parallel to halite crystal growth faces. Core 88-01; largest fluid inclusion is 30 microns in diameter. D, Large, isolated, irregularly shaped fluid inclusion in clear diagenetic halite. Core 88-01; fluid-inclusion diameter is 100 microns. E, Photograph of displacive halite crystal plucked from mud layer 2 m below the surface of the Qarhan Salt Plain. Light bands are relatively rich in fluid inclusions. Crystal diameter is 1.5 cm.

tion of clear, fluid inclusion–poor halite, which partly cements the primary halite crystal framework. Clear, anhedral crystals of carnallite may also form a void filling cement in halite crystal frameworks (Casas et al., 1992). Vugs and fractures in halite layers are filled with large clear crystals of halite (Fig. 3A), and in the top 13 m, with large crystals of carnallite.

Mud layers most commonly consist of massive brown siliciclastic mud, silt, and sand (Chen and Bowler, 1985). In places, faint flat lamination is visible (millimeter to centimeter scale), but more commonly, there are no sedimentary structures. The cores also contain nondisrupted flat-laminated muds (Fig. 2), which consist of siliciclastic or gypsum crystal sand (fine grained) and silt, brown siliciclastic mud, and black organic/sulfide-rich mud. X-ray diffraction analyses show that sand and silt grains are quartz, feldspar, and gypsum; clays are composed of illite and chlorite with minor smectite and kaolinite. Low Mg-calcite is present in small amounts (Chen and Bowler, 1985).

Saline minerals (halite, and in the top 13 m, carnallite) occur in mud layers as displacively grown crystals. Displacive halite crystals are randomly oriented, millimeter- to centimeter-size clear cubes with incorporated mud (Fig. 3B). In some layers, displacive halite is more abundant than mud. Carnallite occurs as centimeter-size clear anhedral crystals. Small crystals of lenticular gypsum, less than 1 cm in diameter, are dispersed in mud layers at depths below 25 m. It is probable that early diagenetic growth of saline minerals has disrupted the muddy host sediment to produce its massive texture.

Several types of fluid inclusions were recognized in crystals of halite:

1. Primary fluid inclusions (less than 1 micron to rarely about 50 microns or greater in diameter) trapped in dense bands parallel to crystal growth faces of chevron and cumulate halite (Fig. 3C). These fluid inclusions are samples of the lake brines from which the primary halite crystallized at the brine bottom (chevron halite) or at the air-water interface (cumulate halite).

2. Primary fluid inclusions in diagenetic halite cement and displacive crystals that occur as isolated negative crystals or irregularly shaped inclusions. They are larger (commonly several tens of microns in diameter) but far less densely packed than fluid inclusions in chevron/cumulate halite (Fig. 3D). These fluid inclusions are samples of the pore fluid brines from which the diagenetic halite formed.

3. Primary fluid inclusions in diagenetic halite cement and displacive crystals that occur as faintly banded arrays parallel to halite crystal faces (Fig. 3E). In this case petrographic features, such as incorporated mud and random orientation of halite crystals, help distinguish this type of fluid inclusion banded halite as diagenetic in origin.

This study reports exclusively on analysis of primary fluid inclusions in bands parallel to halite crystal growth faces (i.e., Fig. 3C) because they alone carry a record of surface brines in the Qaidam Basin over the 54-kyr interval of evaporative accumulation. Halite samples with fluid inclusions were broken into millimeter-size cleavage fragments and placed in a Fluid Inc.–adapted USGS gas-flow heating and freezing stage mounted to a microscope. Fluid inclusions ranged in size from less than 1 micron to about 100 microns in diameter; those between 10 and 20 microns were ideal for observation during freezing-heating runs at magnifications of 400×. Samples were rapidly supercooled to temperatures of –100°C, and then heated at the rate of about 2 to 3°C per minute or more slowly near phase disappearance temperatures. Most fluid inclusions froze during cooling or during heating between temperatures of –100°C and –70°C. The texture of frozen inclusions was a cryptocrystalline "glassy" mass or a coarser, granular mass of submicron size crystals quite similar in appearance to the frozen inclusions described by Davis et al. (1990) from laboratory-grown halites. Frozen fluid inclusions from the Qaidam Basin halites showed complex melting behaviors. The final melting temperature of ice and the final melting temperature of hydrohalite ($NaCl \cdot 2H_2O$) were readily observed petrographically and were compared to phase equilibrium relations of these minerals reported by Spencer et al. (1990a) and the experimental study of Davis et al. (1990). The activity of H_2O in fluid inclusion brines was determined from the final melting temperature of hydrohalite and the data of Spencer et al. (1990a). The final melting temperature of hydrohalite was also used to quantify the total molality of Ca plus Mg in fluid inclusion brines because only $CaCl_2$ and $MgCl_2$ salts significantly depress the melting temperatures of ice or hydrohalite in halite saturated brines in the system Na-K-Mg-Ca-Cl-SO_4 (Table 1; see Davis et al., 1990, their Figs. 3, 5, and 8). Individual molalities of Mg and Ca (dissolution temperatures of hydrohalite give $m_{Mg} + m_{Ca}$), were determined by crush-leach tests and chemical analysis of the leached inclusion brines via atomic absorption spectrophotometry or DCP-atomic emission spectrophotometry. Leachate analyses gave the molar ratio of Mg/Ca in the fluid inclusion brines. With knowledge of the melting temperature of hydrohalite (total Mg + Ca), the Mg/Ca ratio of fluid inclusion brines, the equilibrium constant of halite, and charge balance, the major-element composition (Mg-Ca-Na-Cl) of the fluid inclusions could be determined using the low-temperature ion interaction solution model of Spencer et al. (1990a; Table 1). The dominant ions in all fluid inclusion brines are Mg, Na, and Cl; the calcium content is always subordinate. Figure 4 shows a_{H_2O} and m_{Mg} in fluid inclusions versus depth/age, ignoring the small amount of Ca in fluid inclusions. Mg is the most conservative major element in the fluid inclusion brines; therefore m_{Mg}, along with a_{H_2O}, serve as general concentration gages. Relatively concentrated fluid inclusion brines with high m_{Mg} and low a_{H_2O} occur in halite from depths of 0 to 4 m, 9 to 14 m, and 42 to 43 m. Fluid inclusions with lower m_{Mg} (generally <1 molal) are present in halite at 4 to 8 m, 22 to 28 m, and 35 to 39 m.

Daughter crystals of carnallite were identified in some fluid inclusions in modern halite from the margins of Dabusun Lake. The dissolution temperature of carnallite provides infor-

TABLE 1. MAJOR-ELEMENT CHEMISTRY OF PRIMARY FLUID INCLUSIONS IN CHEVRON/CUMULATE HALITE, CORE 88-01 AND CORE 89-04*

Depth (m)	Final Melt Hydrohalite (°C)	$^{m}Mg2+$	$^{m}Na+$	$^{m}Cl-$	$^{a}H_2O$
Core 88-01					
0.02	-8.0	2.32	1.90	6.54	0.69
0.99	-6.4	2.11	2.25	6.47	0.70
9.06	-7.9	2.31	1.92	6.54	0.69
9.85	-8.1	2.33	1.89	6.54	0.69
10.23	-2.2	1.09	3.84	6.02	0.74
10.39	-2.5	1.20	3.68	6.07	0.73
11.84	-2.1	1.06	3.89	6.00	0.74
12.20	Did not freeze	2.83 to 3.67	1.20 to ≤0.55	6.86 to 8.32	0.65 to 0.56
13.20	-8.9	2.41	1.75	6.57	0.68
14.26	-2.0	1.02	3.95	5.98	0.74
22.93	-1.2	0.70	4.42	5.81	0.75
24.31	-1.2	0.70	4.42	5.81	0.75
24.48	-0.9	0.56	4.61	5.74	0.75
28.37	-1.7	0.90	4.12	5.92	0.74
36.89	-4.0	1.64	3.01	6.28	0.72
37.01	-1.8	0.94	4.06	5.94	0.74
37.29	-1.3	0.74	4.35	5.83	0.75
37.62	-0.6	0.42	5.81	5.66	0.75
37.62	-1.0	0.61	4.54	5.76	0.75
37.76	-1.0	0.61	4.54	5.76	0.75
37.93	-1.5	0.82	4.23	5.88	0.75
38.28	-1.3	0.74	4.35	5.83	0.75
38.28	-1.7	0.90	4.12	5.92	0.74
38.51	-1.5	0.82	3.23	5.88	0.75
38.62	-1.5	0.82	4.23	5.88	0.75
41.81	-0.6	0.42	4.81	5.66	0.75
41.84	-1.2	0.70	4.42	5.81	0.75
42.06	-0.6	0.42	4.81	5.66	0.75
42.65	-1.3	0.74	4.35	5.83	0.75
42.91	-8.5	2.37	1.82	6.56	0.68
43.21	-2.3	1.13	3.78	6.04	0.74
Core 89-04					
1.81	Did not freeze	2.83 to3.67	1.20 to ≤0.55	6.86 to 8.32	0.65 to 0.56
1.95	-4.0	1.64	3.01	6.28	0.72
2.00	-2.9	1.33	3.48	6.14	0.73
2.07	-6.9	2.18	2.13	6.50	0.70
3.93	-3.5	1.51	3.21	6.22	0.72
4.04	-4.2	1.69	2.93	6.30	0.72
6.72	-2.5	1.20	3.68	6.07	0.73
7.03	-2.6	1.23	3.63	6.09	0.73
Core 89-04 (continued)					
7.86	-3.2	1.42	3.34	6.18	0.73
8.95	-2.5	1.20	3.68	6.07	0.73
9.36	-1.0	0.61	4.54	5.76	0.75
10.13	-3.0	1.36	3.44	6.15	0.73
11.43	-8.8	2.40	1.77	6.57	0.68
11.43	-5.5	1.96	2.49	6.42	0.71
13.29	-1.3	0.74	4.35	5.83	0.75
13.92	-6.9	2.18	2.13	6.50	0.70
14.63	-3.4	1.48	3.25	6.21	0.73
22.80	-1.3	0.74	4.35	5.83	0.75
22.90	-1.2	0.70	4.42	5.81	0.75
22.97	-1.0	0.61	4.54	5.76	0.75
24.16	-0.9	0.56	4.61	5.74	0.75
24.21	-1.0	0.61	4.54	5.76	0.75
25.73	-1.1	0.65	4.48	5.79	0.75
25.80	-1.5	0.82	4.23	5.88	0.75
28.03	-1.7	0.90	4.12	5.92	0.74
28.25	-1.0	0.61	4.54	5.76	0.75
34.98	-2.3	1.13	3.78	6.04	0.74
36.53	-1.0	0.61	4.54	5.76	0.75
36.61	-1.2	0.70	4.42	5.81	0.75
36.67	-0.7	0.47	4.74	5.69	0.75
36.73	-1.0	0.61	4.54	5.76	0.75
36.83	-1.7	0.90	4.12	5.92	0.74
37.70	-0.8	0.52	4.68	5.71	0.75
37.73	-1.4	0.78	4.29	5.86	0.75
37.77	-1.8	0.94	4.06	5.94	0.74
37.92	-1.8	0.94	4.06	5.94	0.74
38.02	-0.9	0.56	4.61	5.74	0.75
38.07	-0.8	0.52	4.68	5.71	0.75
38.14	-1.3	0.74	4.35	5.83	0.75
42.60	-3.8	1.59	3.09	6.26	0.72
42.74	-3.5	1.51	3.21	6.22	0.72
Dabusun Lake					
Surface	7.0†	3.66	≤0.55	8.32	0.56
Surface	Did not freeze	2.83 to 3.67	1.20 to ≤0.55	6.86 to 8.32	0.65 to 0.56

*Concentrations in molality.
†Final dissolution temperature of carnallite daughter crystals.

mation on the molalities of K and Mg in fluid inclusion brines. Assuming constant maximum molality of K (1 molal), m_{Mg} at carnallite saturation was calculated (Table 1 and Fig. 4). Some fluid inclusions did not freeze or nucleate daughter crystals even after prolonged cooling to temperatures below −100°C. These fluid inclusions (modern surface halites, and halite from depths of 1.8 m and 12.2 m) probably contain concentrated brines, based on experimental evidence from the Na-Mg-Cl salt system (Davis et al., 1990). Fluid inclusions that did not freeze, plotted on Figure 4, are assumed to contain higher m_{Mg} than fluid inclusions that did freeze, but m_{Mg} below carnallite saturation.

Isotopes of hydrogen and oxygen from fluid inclusion waters in halite from core 89-04 were analyzed at the Department of Physics, University of Calgary. Samples of chevron or cumulate halite with primary fluid inclusions were selected for analysis based on petrographic examination of small cleavage fragments. A modification of the vacuum volatilization method of Knauth and Kumar (1981) and Knauth and Beeunas (1986) was used to release the fluid inclusion

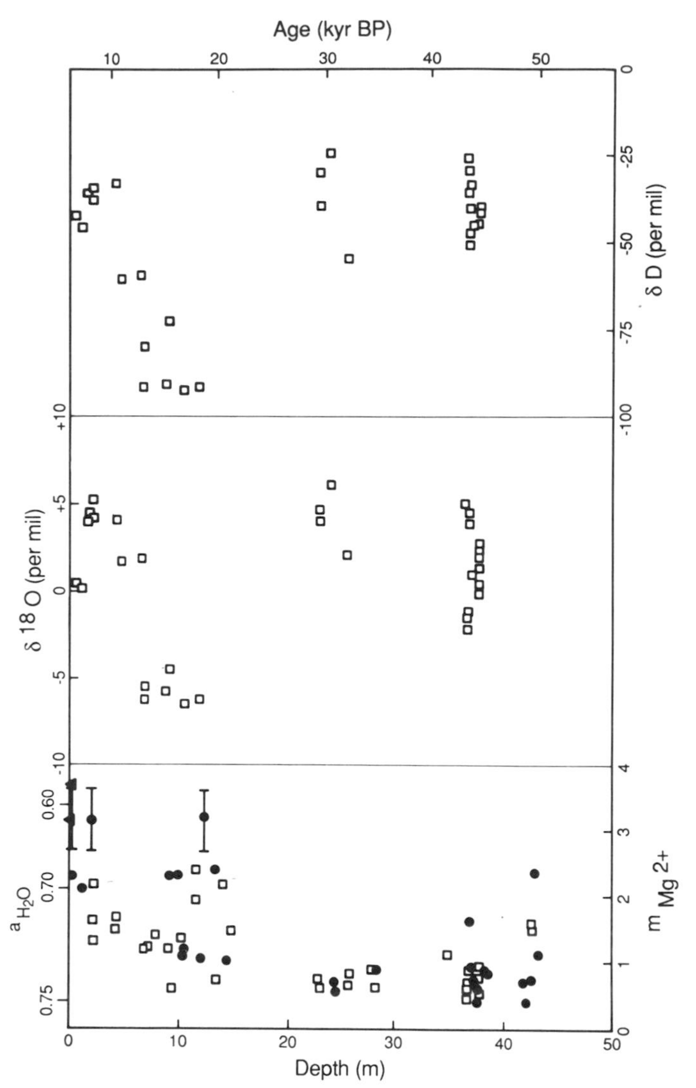

Figure 4. The activity of H_2O-molality of Mg, $\delta^{18}O$ (per mil), and δD (per mil) in primary fluid inclusions from chevron/cumulate halite crystals (solid circles, core 88-01; open squares, core 89-04) plotted versus depth/time. These data provide a record of the composition of surface brines in the Qaidam Basin for the past 50 kyr B.P. Brackets for three values of a_{H_2O}-m_{Mg} represent the possible range of a_{H_2O}-m_{Mg} for fluid inclusions that did not freeze (see Table 1). Two triangles on left side of the a_{H_2O}-m_{Mg} plot represent fluid inclusions in modern halite from Dabusun Lake (one did not freeze and the other had carnallite daughter crystals).

waters (Spencer et al., 1990c; Yang, 1993). Small samples of halite (300 to 600 mg) were placed in a glass tube following cleaning in acetone and ethanol, and removal of adsorbed waters under vacuum. The glass tube was then heated under vacuum until the halite melted (about 950°C), resulting in the complete release of fluid inclusion waters. The glass tube containing the sample of fluid inclusion water was placed in a dry ice–acetone bath and evacuated to remove noncondensed gases. The dry ice–acetone bath was removed, and the glass tube was heated to vaporize the water. The volume of water released from fluid inclusions was between 4 and 10 microliters, based on comparison of the pressure of released fluid inclusion waters with the pressures of known volumes of water. One split of the released fluid inclusion water was reacted with 200 mg zinc at 500°C for one to two hours to produce H_2, following the methods of Friedman (1953) and Coleman et al. (1982). The hydrogen isotope composition of the gas was then analyzed via isotope ratio mass spectrometry (Table 2A; Fig. 4). This technique has been successfully applied to the measurement of δD in fluid inclusions in quartz by Coleman et al. (1982).

A second split of the released fluid inclusion water was used for the determination of oxygen-isotope compositions. This sample was reacted with 100-mg guanidine hydrochloride at 260 C° to produce CO_2, after the methods of Boyer et al (1961) and Dugan et al. (1985). The CO_2 was then analyzed on the mass spectrometer (Table 2A; Fig. 4).

The isotope data from core 89-04 show the following trends with depth: δD and $\delta^{18}O$ in fluid inclusions in halite are high in the upper 6.9 m, lower between depths of 6.9 m and 12.2 m, and high again at depths below 23 m (Fig. 4).

A series of tests documented that the above methods gave reproducible results with no significant fractionation (Yang, 1993). A standard laboratory water and that water with dissolved NaCl or $MgCl_2$ were analyzed using the methods described. The results show that no significant isotope fractionation occurred, even at high concentrations of $MgCl_2$ (up to 30 wt. %). Yang (1993) has also measured δD and $\delta^{18}O$ of natural brines from the Dabusun Lake area of the Qaidam Basin (Table 2B; and inset on Fig. 5), and compared the results with the isotopic composition of fluid inclusions in modern halites that crystallized from those brines. Modern brines show a linear increase in δD and $\delta^{18}O$ as a function of salinity (Yang, 1993). The evaporation path for modern Dabusun Lake brines was determined from the average isotopic composition of modern Golmud River water ($\delta^{18}O$ = –9.425; δD = –69.1), the composition of brines from Dabusun Lake, and the composition of the most concentrated brines from the carnallite flats that fringe Dabusun Lake (Table 2B; and inset on Fig. 5). The Golmud River is the major source of water to Dabusun Lake (Fig. 1). Therefore, dilute Golmud River inflow, Dabusun Lake brines, and brines from the marginal carnallite flats together define an evaporation path with respect to δD and $\delta^{18}O$. The evaporation path of modern waters is linear: $\delta D = 2.6\delta^{18}O - 43.5$ (Fig. 5). Fluid inclusions in modern halites have similar isotopic compositions as modern brines and also lie on the evaporation path (Yang, 1993). These results indicate that primary fluid inclusions in halite record the true isotopic composition of the surface brines from which they precipitate.

Fluid inclusions in halite from core 89-04 have isotopic compositions that fall along the modern evaporation path in the upper 6.9 m and at depths below 23 m; the isotopic compositions of six fluid inclusions from depths of 6.9 m to 12.2 m lie well below the modern evaporation path (Table 2A; and Fig. 5).

TABLE 2A. STABLE ISOTOPES FROM PRIMARY FLUID INCLUSIONS IN CHEVRON/CUMULATE HALITE, CORE 89-04 (relative to SMOW, per mil)

	Depth (m)	δD (‰)	$\delta^{18}O$ (‰)
Core 89-04	0.57	-42.1	0.45
	0.87	-45.5	0.10
	1.81	-35.4	4.10
	1.91	-36.4	4.40
	1.95	-34.6	4.28
	2.07	-37.5	5.20
	4.15	-32.8	4.07
	4.33	-61.0	1.62
	6.61	-59.1	1.90
	6.88	-91.4	-6.19
	7.03	-80.2	-5.49
	8.95	-90.6	-5.73
	9.36	-72.4	-4.55
	10.67	-92.4	-6.53
	12.19	-91.6	-6.28
	22.92	-29.4	4.63
	23.03	-38.9	3.99
	24.14	-23.8	6.14
	25.58	-54.5	2.09
	36.52	-25.7	5.01
	36.62	-29.6	4.73
	36.68	-35.0	4.64
	36.71	-33.6	3.86
	36.74	-47.8	-2.09
	36.86	-40.1	-1.36
	36.89	-50.6	-1.53
	37.32	-44.9	1.05
	37.62	-42.4	1.30
	37.67	-41.6	-0.14
	37.90	-43.4	0.33
	37.92	-41.5	2.20
	37.95	-41.4	2.44
	37.97	....	2.63

TABLE 2B. STABLE ISOTOPES IN MODERN SURFACE WATERS QARHAN AREA, QAIDAM BASIN (relative to SMOW, per mil)

	δD (‰)	$\delta^{18}O$ (‰)
Golmud River	-68.1	-9.59
	-70.2	-9.26
Average	-69.1	-9.425
Dabusun Lake	-39.5	0.48
	-38.2	2.33
	-40.5	1.23
	-46.5	0.32
Halite/Carnallite flats - Dabusun Lake margin:	-26.7	5.41
	-29.4	5.81
	-32.5	3.75
	-28.1	6.67
	-29.0	6.27
	-29.1	6.37
	-34.1	2.72
	-31.6	5.01
	-35.3	2.84
	-31.9	4.28
	-32.4	3.87
	-32.2	5.08
	-31.6	5.02
	-31.3	4.54
	-33.1	4.15
	-32.0	4.59
	-30.9	4.66
	-34.1	3.62
	-33.6	4.11
	-35.4	2.37
	-37.7	1.63
	-34.1	3.05
	-36.8	2.01
	-46.7	-1.58

Some fluid inclusions in cumulate halite crystals contain a gas in addition to brine. This gas is probably air trapped during crystallization of halite at the air-brine interface. Preliminary analysis of atmospheric gas in fluid inclusions has been done by crushing halite crystals under vacuum at room temperature, followed by analysis via quadrupole mass spectrometry. Initial determinations on 7 samples of modern halite and 17 Holocene/late Pleistocene halite samples are plotted on Figure 6 as the ratio of the quadrupole mass spectrometer ion current at mass 44 (CO_2) to that at mass 28 (N_2), relative to the average values of 7 modern halite samples. Modern halites have average CO_2/N_2 close to the modern atmospheric ratio. The shift to lower atmospheric CO_2 in samples between 7 and 12.2 m (13 to 18 kyr B.P.) is similar in magnitude to the shift observed from polar ice core data (Fig. 6). Some of the samples with anomalously high CO_2 (40 to 50 kyr B.P.) relative to the polar ice cores probably formed during early diagenesis and may contain CO_2 derived from organic matter. Horita (1990) also found relatively high values of CO_2 in some fluid inclusions in halite from Searles Lake, California, which, based on measurements of $\delta^{13}C$, were interpreted to have formed from organic matter. Lower values for each sample interval, however, agree well with the polar ice core record and probably represent atmospheric gas (Fig. 6).

DISCUSSION

Borehole cores through evaporites are used to interpret paleoclimates and brine evolution in the Qaidam Basin over the past 50 kyr B.P. based on (1) sedimentology/mineralogy, (2) major-element chemistry of fluid inclusions in primary chevron/cumulate halite, (3) stable isotopes (δD and $\delta^{18}O$) in fluid inclusions in primary halite, and (4) preliminary data on the relative CO_2/N_2 in fluid inclusions from cumulate halites that contain trapped air.

Interpretations of relative aridity in the Qaidam Basin rely on the salinity of fluid inclusions in halite and on the presence or absence of evaporites. Fluid inclusion salinities, expressed

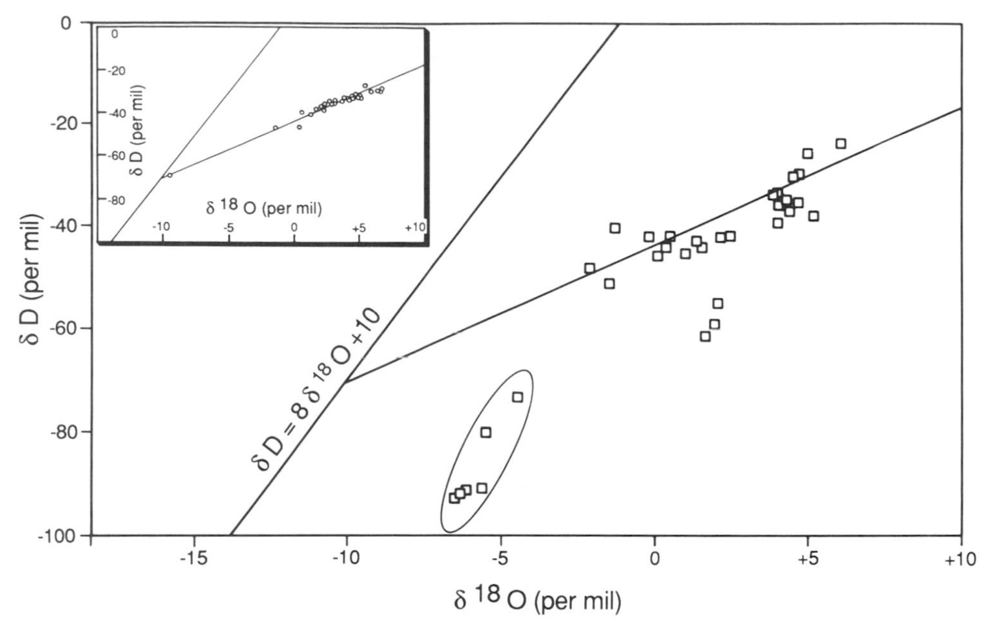

Figure 5. Stable isotope composition of 33 primary fluid inclusion brines from chevron and cumulate halite crystals in core 89-04. Inset gives data for 31 analyses of modern river water, modern lake brines, and modern salt flat brines (from Zhang, 1987; Zhang et al., 1990; Yang, 1993) that define the evaporation path ($\delta D = 2.6\delta^{18}O - 43.5$). Yang (1993) has shown that fluid inclusions in modern halite lie along the evaporation path. Most fluid inclusions (24) from core 89-04 also lie along the modern evaporation path. However, six fluid inclusion brines (circled) from depths of 6.9 to 12.2 m (13 to 18.5 kyr B.P.) fall well below the modern evaporation path. The $\delta^{18}O$ and δD of each fluid inclusion brine is projected parallel to the modern evaporation path onto the meteoric water line ($\delta D = 8\ \delta^{18}O + 10$), to arrive at the isotopic composition of dilute inflow waters through time, shown on Figure 6.

in terms of activity of H_2O and m_{Mg} (a conservative component), reflect changes in the ratio of inflow to evaporation and therefore record relative basin aridity. Stable isotopes of hydrogen and oxygen in fluid inclusion waters in halite are used for the interpretation of surface paleotemperatures. The relative aridity and regional paleotemperatures in the Qaidam Basin may have varied independently of one another.

Relative basin aridity

The Qaidam Basin is a large topographically closed basin in which the volume and salinity of lake waters should reflect the ratio of inflow to evaporation. Elevated shoreline deposits with ostracods and gastropods (carbon-14 dates of 28.6 kyr B.P. and 38.6 kyr B.P.) in the eastern Qaidam Basin document late Pleistocene freshwater lakes (Chen and Bowler, 1986). There is no evidence for loss of basin waters via spillover. Therefore we assume that the Qaidam basin has been hydrologically closed over the last 50 kyr, except for the untested possibility of groundwater leakage (Wood and Sanford, 1990).

Several cored intervals contain mud without any primary layered halite. Some of this mud is finely laminated and undisrupted by desiccation or growth of saline minerals (Fig. 2). These layers represent stages when surface waters were undersaturated with respect to halite and are interpreted to be wetter periods of relatively high inflow to evaporation. Such relatively wet conditions existed from 21 to 29 kyr B.P., 35 to 43 kyr B.P., 45 to 49 kyr B.P., and from 50 kyr B.P. to the base of our record (Fig. 6).

Primary fluid inclusions in chevron and cumulate halite crystals preserve a record of fluctuations in the major-element chemistry of surface brines in the Qaidam Basin through time. The activity of H_2O and the magnesium content of these fluid inclusions record the degree of evaporative concentration the brines have undergone and are used to distinguish relative basin aridity varying from arid (a_{H_2O} of 0.75 or m_{Mg} <1) to extremely arid ($a_{H_2O} < 0.67$, $m_{Mg} > 2.5$, or carnallite daughter crystals).

Modern open lake brines of Dabusun Lake ($a_{H_2O} = 0.60$ to 0.67; $m_{Mg} = 2.4$ to 3.3) are more concentrated than any surface waters in the Qarhan area over the past 54 kyr B.P. (Fig. 6). Carnallite daughter crystals and fluid inclusions that did not freeze, both of which indicate very high salinities, are unique to fluid inclusions from primary halite of modern Dabusun Lake and the surrounding salt flats. Therefore, the modern appears to be the driest climate of the 50-kyr-B.P. record.

The subsurface halite from 0 to 14 m contains fluid inclusions with relatively low a_{H_2O}/high m_{Mg}, which indicates intense evaporative concentration and basin aridity over the last 20 kyr B.P. Elevated salinities in fluid inclusions from primary halite near the base of the cores (42 to 43 m; Figs. 4 and 6),

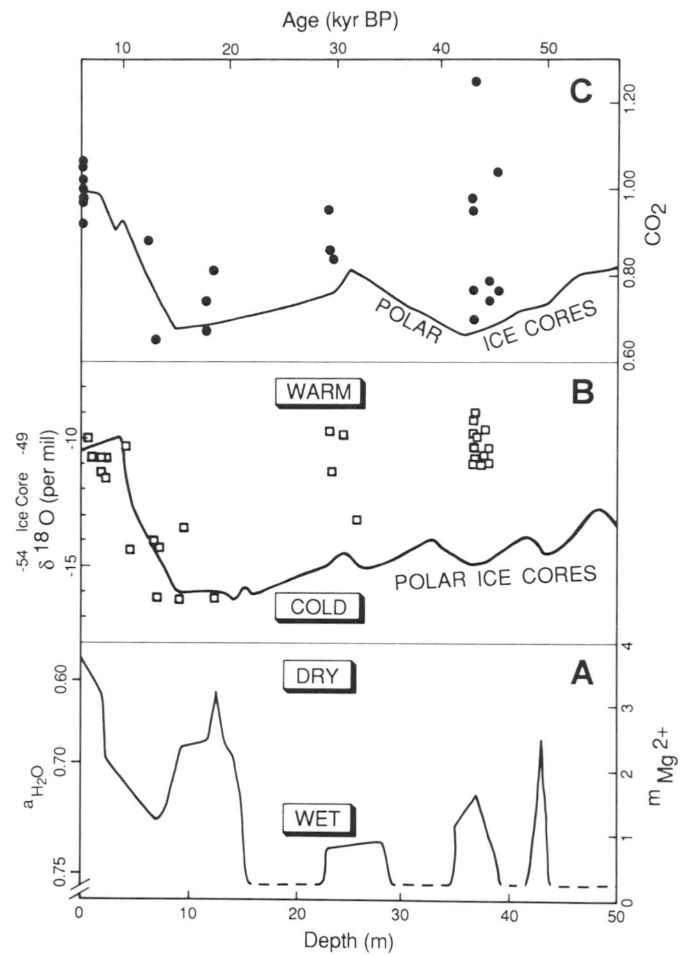

Figure 6. The composition of surface brines preserved in primary fluid inclusions in halite is used to define climate variation in the Qaidam Basin. A, Relative aridity is based on activity of H_2O and major-element concentration of fluid inclusion brines, expressed as m_{Mg}, during periods of salt accumulation. Wetter conditions, shown as flat, dashed lines with high a_{H2O} and low m_{Mg}, are inferred for intervals with no primary salts. B, Oxygen isotopes (core 89-04) for the last 20 kyr B.P. define a trend similar to that obtained from polar ice cores (solid curve, smoothed, from Lorius et al., 1990). Data from older salt samples deviate significantly from the polar ice core record. C, Preliminary atmospheric CO_2 results (from core 89-04), plotted as the ratio of the quadrupole mass spectrometer ion current as mass 44 (CO_2) to that as mass 28 (N_2), relative to the average value for seven modern samples. The shift to lower atmospheric CO_2 from 18 to 13 kyr B.P. is similar in magnitude to the shift observed from polar ice core data (solid curve, smoothed, from Lorius et al., 1990). Some of the samples with high CO_2 (40 to 50 kyr B.P.) probably include diagenetic CO_2, most likely derived from organic matter. Lower values for each sample interval agree well with the polar ice core record and probably represent atmospheric gas.

must have resulted from a period of extreme aridity about 50 kyr B.P. Fluid inclusions from primary halite with higher a_{H_2O} and lower m_{Mg} occur from 22 to 28 m and 35 to 39 m, indicating arid conditions, less extreme than the modern, existed from 29 to 35 kyr B.P. and 43 to 45 kyr B.P.

Relative surface temperatures

The isotopic composition of fluid inclusions in halite is a function of the isotopic composition of dilute inflow waters and their degree of evaporation. The evaporation component may be removed from the isotopic data by back-projecting the isotopic composition of each fluid inclusion analysis onto the meteoric water line, using a slope determined from the modern evaporation path (Fig. 5). For example, fluid inclusions in halite from a depth of 9.36 m have $\delta D = -72.4$ per mil and $\delta^{18}O = -4.55$ per mil. After projection back to the meteoric water line using the slope of the modern evaporation path ($\delta D = 2.6\delta^{18}O - 43.5$), the "preevaporated" water has a calculated $\delta D = -95.83$ per mil and $\delta^{18}O = -13.23$ per mil. This projection gives the isotopic composition of dilute inflow waters, prior to evaporation, assuming a constant meteoric water line and evaporation path, both of which may have varied some. The results (Fig. 6) are a record of the isotopic composition of precipitation in the high mountains surrounding the Qaidam Basin because Lowenstein et al. (1989) and Spencer et al. (1990b) have shown that these waters are the major source of inflow to modern Dabusun Lake. The calculated oxygen isotopic composition of inflow waters, shown as $\delta^{18}O$ versus depth/age in Figure 6, ranges from near –17 to –8 per mil. This range is similar to that found in ice cores from the Qilian Mountains, which border the Qaidam Basin to the north (for example, oxygen isotopes from –14 to –8 per mil over the last 40 kyr B.P., see Thompson et al., 1989).

The isotopic composition of precipitation is produced by complex processes of fractionation during evaporation of seawater, transport of air masses inland, and precipitation as rain and snow. The isotopic composition of precipitation decreases (becomes lighter) with transport distance; isotope fractionation is also influenced by temperature during evaporation and precipitation. The isotopic variation of the inflow waters to the Qaidam Basin, deduced from fluid inclusions, is interpreted to reflect changes in the temperature of precipitation, which may be correlated with surface temperatures (see discussions in Harmon et al., 1979; Winograd et al., 1988). Variations in the isotopic composition of fluid inclusion waters, however, may also result from changes in atmospheric circulation patterns, sources of moisture, and seasonality of precipitation.

The isotopic composition of fluid inclusion waters in primary halite changes dramatically over the cored interval. Fluid inclusions from depths of 6.9 to12.2 m have exceptionally low oxygen and hydrogen isotopes relative to all other primary fluid inclusions (8 to 10 per mil and 30 to 55 per mil lower, respectively; Figs. 4 and 6), which are interpreted to reflect regionally cooler climate from 18.5 to 13 kyr B.P. Low values of δD and $\delta^{18}O$ in fluid inclusions in halite from the Qaidam Basin during the last glacial period, together with abnormally low values of δD and $\delta^{18}O$ in groundwaters from the southern Great Basin thought to represent late Pleistocene recharge (Benson and Klieforth, 1989), support the interpretation that isotopes of hydrogen and oxygen are lower during colder climate periods. This evidence suggests that the isotopic composition of fluid inclusion waters in evaporites has great potential for documenting glacial-interglacial climate signals.

SUMMARY

A record of brine evolution and paleoclimate may be obtained from major elements, stable isotopes, and CO_2 content of primary fluid inclusions in crystals of halite that contain trapped surface brines and in some cases, air (Fig. 6). Variation in the major elements and activity of H_2O document changes in the ratio of inflow to evaporation and therefore provide a record of relative basin aridity. Stable isotopes (oxygen and hydrogen) are influenced by both the isotopic composition of the inflow waters and evaporative concentration. The inflow composition can be obtained by subtracting the effects of evaporation. The isotopic record of the inflow waters can then document variations in surface temperatures with time. Relative CO_2/N_2 in fluid inclusions from cumulate halite preserve a record of atmospheric gas chemistry.

The Qaidam Basin was relatively wet with lacustrine deposits lacking evaporites from 50 kyr B.P. to the base of the core record. Conditions of extreme aridity and salt deposition occurred at about 50 kyr B.P. Wetter conditions existed from 50 to 45 kyr B.P. because no salts were deposited. Halite accumulation between 45 to 41 kyr B.P., marked by warmer, more arid conditions, was followed by a return to a wetter climate between 41 and 35 kyr B.P. (no salt). Another cycle of warmer, arid conditions (35 to 29 kyr B.P.) was followed by relatively wet conditions, and nonsaline lacustrine deposits between 29 and 21 kyr B.P. Salt precipitated under very dry, but cold conditions between 20 and 13 kyr B.P. Warming occurred about 13 kyr B.P. and warm arid conditions have persisted and intensified to the present.

Comparison with other climate records

The climate in the Qaidam Basin from 50 to 20 kyr B.P. was wetter than the modern, but similar with respect to temperature ($\delta^{18}O$); $\delta^{18}O$ and δD generally decrease over this period, and relative CO_2 is below the modern values. Ice cores from Antarctica also have generally decreasing values of $\delta^{18}O$, δD, and CO_2 for the same period but absolute values indicate colder than modern temperatures (Lorius et al., 1985, 1990; Barnola et al., 1987; Jouzel et al., 1987). Such differences between the Qaidam salt cores and the Antarctic ice cores may reflect different atmospheric circulation patterns over the Qinghai-Tibet Plateau during the Pleistocene, or differences in the magnitude of midlatitude and polar temperature variations.

The salt cores from the Qaidam Basin reveal dramatic shifts in $\delta^{18}O$, δD, atmospheric CO_2, and salinity during the last 20 kyr B.P., which coincide with those documented in the nearby Qilian Mountain ice cores (Thompson et al., 1989) and the polar ice cores (Lorius et al., 1990). Large shifts in the oxygen isotopes in fluid inclusions in halite from −10 per mil in the lower halite layers, to −13 per mil at about 21 kyr B.P., to −17 per mil from 19 to 15 kyr B.P., and a return to values of about −10 per mil at 10 kyr B.P. (Fig. 6) coincide with global cooling during the last glacial (minima in δD, $\delta^{18}O$, and atmospheric CO_2 and CH_4 in Antarctic ice cores; see Jouzel et al., 1987; Lorius et al., 1985; Chappellaz et al., 1990). The low CO_2/N_2 found in cumulate halite from depths of 7 to 12 m (13 to 18 kyr B.P.) adds support for the existence of cooler glacial conditions, when atmospheric CO_2 levels were lower than modern values. Colder, glacial climates are also indicated by the general 2 per mil lowering of $\delta^{18}O$ in the Qilian Mountain ice cores for the period 30 to 12 kyr B.P. (Thompson et al., 1989). The high salinity and low activity of H_2O of fluid inclusions in halite also establishes that the ratio of inflow to evaporation in the Qaidam Basin was low during this glacial period, indicating hyperarid climate conditions. This is quite unexpected considering that pluvial lakes reaching maximum depths between 29 and 13.5 kyr B.P., developed in western North America during the last glacial (Spencer et al., 1984; Benson et al., 1990), and mountain snowlines in North America, South America, and New Zealand were at much lower elevations than they are today (Broecker and Denton, 1989). Although there is no consensus, most geoscientists believe that the extent of glacial advance on the Tibetan Plateau was relatively minor during the last glacial (Zheng, 1989; Burbank and Kang, 1991). The salt core record suggests that glacial buildup on the Qinghai-Tibet Plateau was limited by the arid climate.

Finally, the last glacial-interglacial transition (15 to 10 kyr B.P.) and the Holocene in the Qaidam Basin salt cores are marked by increased $\delta^{18}O$, δD, and atmospheric CO_2 similar in magnitude to that observed in the Antarctic ice cores as well as the ice cores from the Qilian Mountains. Taken together, the data suggest that the Holocene climate of the Qaidam Basin has been dry and warm, and has culminated with the development of the world's largest modern potash evaporite deposit.

ACKNOWLEDGMENTS

We are grateful for the generous field and logistical support provided by the Institute of Salt Lakes in Xining. Reviewers Kinga Revesz, Blair Jones, and Warren Wood had many thoughtful suggestions that were used in revising the manuscript. We thank the Petroleum Research Fund, administered by the American Chemical Society (PRF-18652-G2 and 21130-AC2) and the National Science Foundation (EAR8816111) for support of this research.

REFERENCES CITED

Barnola, J. M., Raynaud, D., Korotkevich, Y. S., and Lorius, C., 1987, Vostok ice core provides 160,000-year record of atmospheric CO_2: Nature, v. 329, p. 408–414.

Benson, L. V., and Klieforth, H., 1989, Stable isotopes in precipitation and groundwater in the Yucca Mountain region, southwest Nevada: Paleoclimatic implications, *in* Peterson, D. H., ed., Aspects of climate variability in the Pacific and the western Americas: American Geophysical Union, Geophysical Monograph 55, p. 41–59.

Benson, L. V., and 7 others, 1990, Chronology of expansion and contraction of four Great Basin lake systems during the past 35,000 years: Paleogeography, Palaeoclimatology, Palaeoecology, v. 78, p. 241–286.

Bischoff, J. L., Rosenbauer, R. J., and Smith, G. I., 1985, Uranium-series of sediments from Searles Lake: Differences between continental and marine climate records: Science, v. 227, p. 1222–1224.

Bowler, J. M., Huang, Q., Chen, K., Head, M. J., and Yuan, B., 1986, Radiocarbon dating of playa-lake hydrologic changes: Examples from northwestern China and central Australia: Palaeogeography, Palaeoclimatology, Palaeoecology, v. 54, p. 241–260.

Boyer, P. D., Graves, D. J., Suelter, C. H., and Dempsey, M. E., 1961, Simple procedure for conversion of oxygen of orthophosphate or water to carbon dioxide for oxygen-18 determination: Analytical Chemistry, v. 33, p. 1906–1909.

Broecker, W. S., and Denton, G. H., 1989, The role of ocean-atmosphere reorganizations in glacial cycles: Geochimica et Cosmochimica Acta, v. 53, p. 2465–2501.

Burbank, D. W., and Kang, J. C., 1991, Relative dating of Quaternary moraines, Rongbuk Valley, Mount Everest, Tibet: Implications for an ice sheet on the Tibetan Plateau: Quaternary Research, v. 36, p. 1–18.

Casas, E., and Lowenstein, T. K., 1989, Diagenesis of saline pan halite: Comparison of petrographic features of modern, Quaternary, and Permian halite: Journal of Sedimentary Petrology, v. 59, p. 724–739.

Casas, E., Lowenstein, T. K., Spencer, R. J., Zhang, P., 1992, Carnallite mineralization in the nonmarine Qaidam Basin, China: Evidence for the early diagenetic origin of potash evaporites: Journal of Sedimentary Petrology, v. 62, p. 881–898.

Chappellaz, J., Barnola, J. M., Raynaud, D., Korotkevich, Y. S., and Lorius, C., 1990, Ice-core record of atmospheric methane over the past 160,000 years: Nature, v. 345, p. 127–131.

Chen, K., and Bowler, J. M., 1985, Preliminary study of sedimentary characteristics and evolution of palaeoclimate of Qarhan Salt Lake in Qaidam Basin: Scientia Sinica (series B), v. 28, p. 1218–1232.

Chen, K., and Bowler, J. M., 1986, Late Pleistocene evolution of salt lakes in the Qaidam Basin, Qinghai Province, China: Palaeogeography, Palaeoclimatology, Palaeoecology, v. 54, p. 87–104.

Coleman, M. L., Shepherd, T. J., Durham, J. J., Rouse, J. E., and Moore, G. R., 1982, Reduction of water with zinc for hydrogen isotope analysis: Analytical Chemistry, v. 54, p. 993–995.

Davis, D. W., Lowenstein, T. K., and Spencer, R. J., 1990, Melting behavior of fluid inclusions in laboratory-grown halite crystals in the systems $NaCl\text{-}H_2O$, $NaCl\text{-}KCl\text{-}H_2O$, $NaCl\text{-}MgCl_2\text{-}H_2O$, and $NaCl\text{-}CaCl_2\text{-}H_2O$: Geochimica et Cosmochimica Acta, v. 54, p. 591–601.

Dugan, J. P., Jr., and 8 others, 1985, Guanidine hydrochloride method for determination of water oxygen isotope ratios and the oxygen-18 fractionation between carbon dioxide and water at 25°C: Analytical Chemistry, v. 57, p. 1734–1736.

Eugster, H. P., and Hardie, L. A., 1978, Saline lakes, *in* Lerman, A., ed., Lakes: Chemistry, geology, physics: New York, Springer-Verlag, p. 237–293.

Friedman, I., 1953, Deuterium content of natural water and other substances: Geochimica et Cosmochimica Acta, v. 4, p. 89–103.

Hardie, L. A., 1984, Evaporites: marine or non-marine?: American Journal of Science, v. 284, p. 193–240.

Hardie, L. A., Smoot, J. P., and Eugster, H. P., 1978, Saline lakes and their deposits: A sedimentological approach, *in* Matter, A., and Tucker, M. E., eds., Modern and ancient lake sediments: Oxford, England, Blackwell Scientific Publications, International Association of Sedimentologists Special Publication 2, p. 7–41.

Harmon, R. S., Schwarcz, H. P., and O'Neil, J. R., 1979, D/H ratios in speleothem fluid inclusions: A guide to variations in the isotopic composition of meteoric precipitation?: Earth and Planetary Science Letters, v. 42, p. 254–266.

Horita, J., 1990, Stable isotope paleoclimatology of brine inclusions in halite: Modeling and application to Searles Lake California: Geochimica et Cosmochimica Acta, v. 54, p. 2059–2073.

Jannik, N. O., Phillips, F. M., Smith, G. I., and Elmore, D., 1991, A ^{36}Cl chronology of lacustrine sedimentation in the Pleistocene Owens River system: Geological Society of America Bulletin, v. 103, p. 1146–1159.

Jouzel, J., and 6 others, 1987, Vostok ice core: A continuous isotope temperature record over the last climatic cycle (160,000 years): Nature, v. 329, p. 403–408.

Knauth, L. P., and Beeunas, M. A., 1986, Isotope geochemistry of fluid inclusions in Permian halite with implications for the isotopic history of ocean water and the origin of saline formation waters: Geochimica et Cosmochimica Acta, v. 50, p. 419–433.

Knauth, L. P., and Kumar, M. B., 1981, Trace water content of salt in Louisiana salt domes: Science, v. 213, p. 1005–1007.

Lorius, C., and 6 others, 1985, A 150,000-year climatic record from Antarctic ice: Nature, v. 316, p. 591–596.

Lorius, C., Jouzel, J., Raynaud, D., Hansen, J., and Le Treut, H., 1990, The ice core record: Climate sensitivity and future greenhouse warming: Nature, v. 347, p. 139–145.

Lowenstein, T. K., and Hardie, L. A., 1985, Criteria for the recognition of salt-pan evaporites: Sedimentology, v. 32, p. 627–644.

Lowenstein, T. K., Spencer, R. J., and Zhang, P., 1989, Origin of ancient potash evaporites: Clues from the modern nonmarine Qaidam Basin of western China: Science, v. 245, p. 1090–1092.

Manabe, S., and Broccoli, A. J., 1990, Mountains and arid climates of middle latitudes: Science, v. 247, p. 192–195.

Ruddiman, W. F., and Kutzbach, J. E., 1989, Forcing of late Cenozoic Northern Hemisphere climate by plateau uplift in southern Asia and the American west: Journal of Geophysical Research, v. 94, no. D15, p. 18409–18427.

Smith, G. I., 1984, Paleohydrologic regimes in the southwestern Great Basin, 0–3.2 my ago, compared with other long records of "global" climate: Quaternary Research, v. 22, p. 1–17.

Spencer, R. J., and 11 others, 1984, Great Salt Lake, and precursors, Utah: The last 30,000 years: Contributions to Mineralogy and Petrology, v. 86, p. 321–334.

Spencer, R. J., Moller, N., and Weare, J. H., 1990a, The prediction of mineral solubilities in natural waters. A chemical equilibrium model for the $Na\text{-}K\text{-}Ca\text{-}Mg\text{-}Cl\text{-}SO_4\text{-}H_2O$ system at temperatures below 25°C: Geochimica et Cosmochimica Acta, v. 54, p. 575–590.

Spencer, R. J., Lowenstein, T. K., Casas, E., and Zhang, P., 1990b, Origin of potash salts and brines in the Qaidam Basin, China, *in* Spencer, R. J., and Chou, I. M., eds., Fluid-mineral interactions: A tribute to H. P. Eugster: Geochemical Society Special Publication 2,, p. 395–408.

Spencer, R. J., and 7 others, 1990c, Stable isotope analysis of fluid inclusions in halite from the Qarhan salt plain, Qaidam Basin, China: Geological Society of America Abstracts with Programs, v. 22, p. 249.

Thompson, L. G., and 9 others, 1989, Holocene–late Pleistocene climatic ice core records from Qinghai-Tibetan Plateau: Science, v. 246, p. 474–477.

Wang, Q., and Coward, M. P., 1990, The Chaidam Basin (NW China): Formation and hydrocarbon potential: Journal of Petroleum Geology, v. 13, no. 1, p. 93–112.

Winograd, I. J., Szabo, B. J., Coplen, T. B., and Riggs, A. C., 1988, A

250,000-year climatic record from Great Basin vein calcite: Implications for Milankovitch Theory: Science, v. 242, p. 1275–1280.
Wood, W. W., and Sanford, W. E., 1990, Ground-water control of evaporite deposition: Economic Geology, v. 85, p. 1226–1235.
Wu, B., Duan, Z., Guan, Y., and Lian, W., 1986, Deposition of potash-magnesium salts in the Qarhan Playa, Qaidam Basin: Acta Geologica Sinica, v. 60, p. 79–90.
Yang, W., 1993, Improved techniques for stable isotope analyses of microlitre quantities of water and applications to paleoclimate and diagenesis using fluid inclusions in halite and dolomite [Ph.D. thesis]: University of Calgary, 151 p.
Zhang, B., Fan, H., Zhang, P., Lowenstein, T. K., and Spencer, R. J., 1990, Hydrogen and oxygen stable isotope analyses of fluid inclusions in halite in Charhan Salt Lake with geochemical implications: Acta Sedimentologica Sinica, v. 8, p. 3–18.
Zhang, P., 1987, Saline lakes of the Qaidam Basin (in Chinese): Beijing, Publishing House of Science, 150 p.
Zheng, B., 1989, Controversy regarding the existence of large ice sheet on the Qinghai-Xizang (Tibetan) Plateau during the Quaternary Period: Quaternary Research, v. 32, p. 121–123.

Manuscript Accepted by the Society July 2, 1993

Printed in U.S.A.

Geological Society of America
Special Paper 289
1994

The Sudanese buried saline lakes

Ramsis B. Salama
Division of Water Resources, CSIRO, Private Bag, Wembley, 6014 Western Australia

ABSTRACT

The concepts of plate tectonics have transformed the interpretation of the sedimentary basins in Sudan. Improved methods of mapping, extensive geophysical surveys, and widespread drilling have led to a better understanding of the geological processes. With continuous subsidence in the grabens and troughs, and continuous uplift in the flanking areas, hydrological and hydrogeological closed basins were formed in each of the rift systems. Highly saline groundwater bodies that occupy the flowing end of each of the rift basins have been interpreted in the light of the rift basins. A new chemical pattern emerged, which led to the interpretation of these saline bodies as buried saline lakes, sabkhas, or playas. The thick carbonate deposits existing at the faulted boundaries define the possible contact between the fresh and saline water bodies. The widespread presence of kankar nodules in the sediments was a result of continuous efflorescence, leaching, and evaporative process. The saline water bodies are formed through salt leaching and groundwater discharge.

SETTING THE SCENE

Structural controls

The deep lineaments and fault patterns of Sudan follow two main directions (Fig. 1[1]): trending north-northwest (Red Sea trend) and trending east-northeast (Gulf of Aden trend; Salama, 1985a, b; Schull, 1988). Precambrian mobile belts trend northeast and northwest (Vail, 1978; Almond, 1982; Ahmed, 1982), and Palaeozoic(?) sediments occupy northeast-southwest–aligned grabens (Salama, 1985a). Mesozoic continental sediments with northwest palaeotrends were deposited in major depressions also aligned northwest (Salama, 1985a).

Three phases of rifting are recognized (Browne and Fairhead, 1983; Browne et al., 1985; Salama, 1985a, 1993; Schull, 1988). The first phase was thought to have begun in the Jurassic(?)–lower Cretaceous (130 to 160 Ma). No volcanic activity is associated with this rift phase. The second phase, which occurred during the Turonian–late Senonian (Upper Cretaceous), was associated with minor volcanic activity. The third rift phase began in the late Eocene–Oligocene, during which scattered volcanism occurred (Vail, 1978). This was followed by an intracratonic sag phase of very gentle subsidence accompanied by little or no faulting. The sag period was associated with extensive volcanism parallel to the Zalingei folded belt (Salama, 1985b).

The effect of rifting caused the development of a complicated system of grabens having a predominant north-northwest

[1]Figures 1–5 are reproduced from:

Salama, R. B., 1991, The role of Megastructures on the development of salinity in the River Nile Basins, *in* Australian Water Resources Council Conference Series No. 20, Proceedings of the International Conference on Groundwater in large sedimentary basin (Perth, W. Australia, 1990): Canberra, Australian Government Publishing Service, p. 288–297.

Salama, R. B., 1994, The Sudanese buried saline lakes, *in* Rosen, M. R., ed., Paleoclimate and Basin Evolution of Playa Systems: Boulder, Colorado, Geological Society of America Special Paper 289.

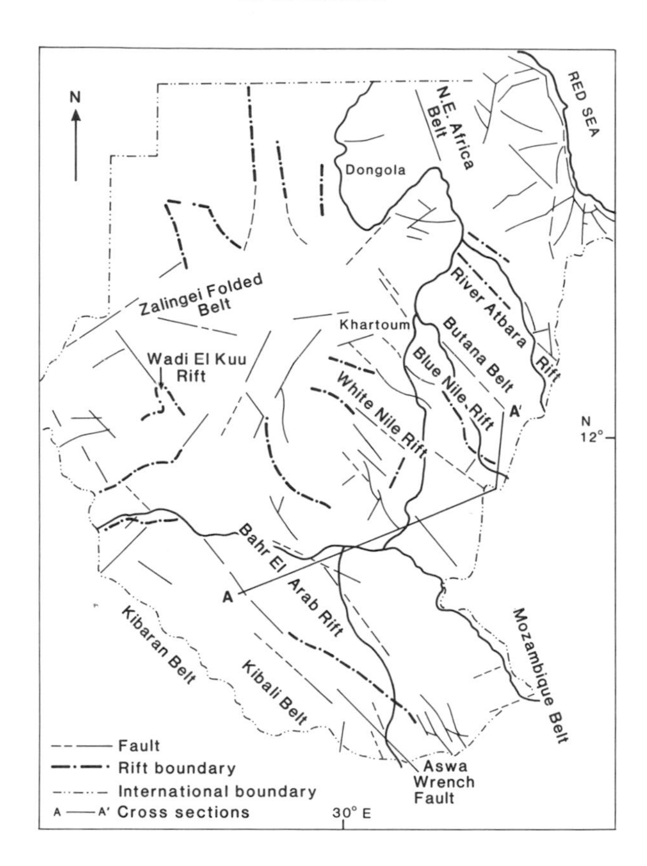

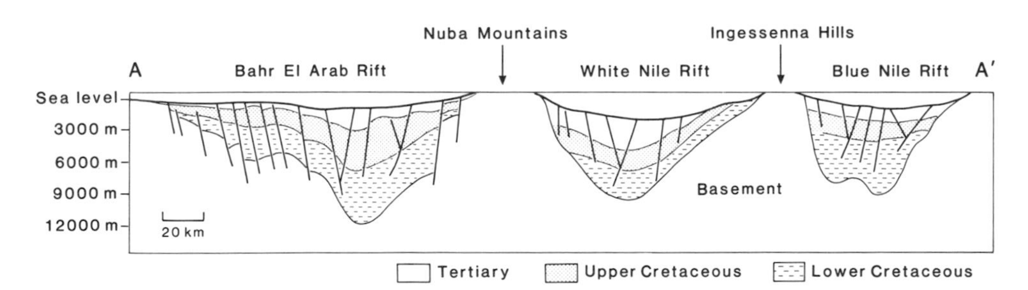

Figure 1. Precambrian mobile belts, major faults, and rift systems in Sudan. Section A–A′ crosses three of the rift systems, showing the normal faults, uplifted mountains, and sedimentary sequence.

direction formed by dip-slip faulting associated with rotated fault blocks, drape folds, and reverse drag folds (Schull, 1988). The Sudanese Cainozoic rift system, forms the largest rift system in Africa, it extends from the eastern borders of Sudan with Ethiopia to the western borders with Central African Republic and Chad. It includes from west to east; Bahr El Arab rift, Wadi El Kuu rift, White Nile rift, Blue Nile rift, and River Atbara rift (Fig. 2). The thick Jurassic(?)-Cretaceous and Tertiary sediments filling the deep grabens (Schull, 1988) were eroded from the following elevated blocks: Jebel Marra, Nuba Mountains, Ingessena Hills, Darfur Dome, and the Nile Congo divide (Salama, 1985a, 1987; Schull, 1988).

Hydrological controls

With continuous subsidence in the grabens and troughs, and continuous uplift in the flanking areas, hydrological and hydrogeological closed basins were formed in each of the rift systems as follows: the Sudd in Bahr El Arab rift, the Nuba in the White Nile rift, the Gezira and Soba in the Blue Nile rift, the Atbara and Gash in the Atbara rift (Salama, 1987; Fig. 3).

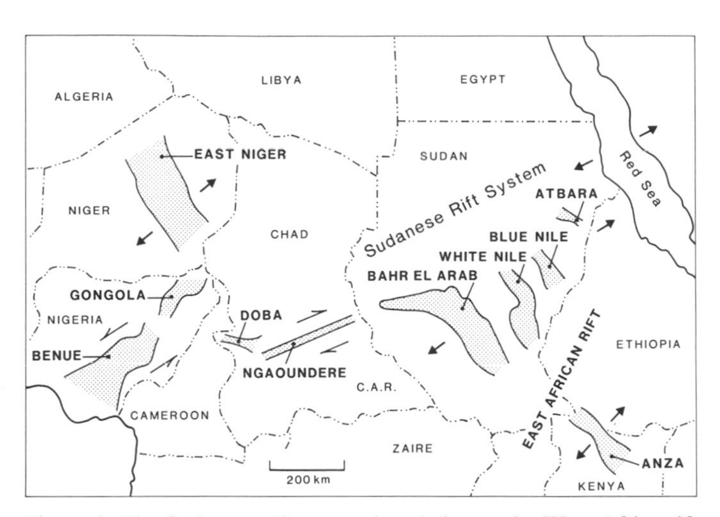

Figure 2. The Sudanese rift system in relation to the West Africa rift system and the East African rift system (stippled areas). Showing major zones of extension (complete arrow) and location of shear zone (half arrow). Modified from Salama, (1985a) and Schull (1988).

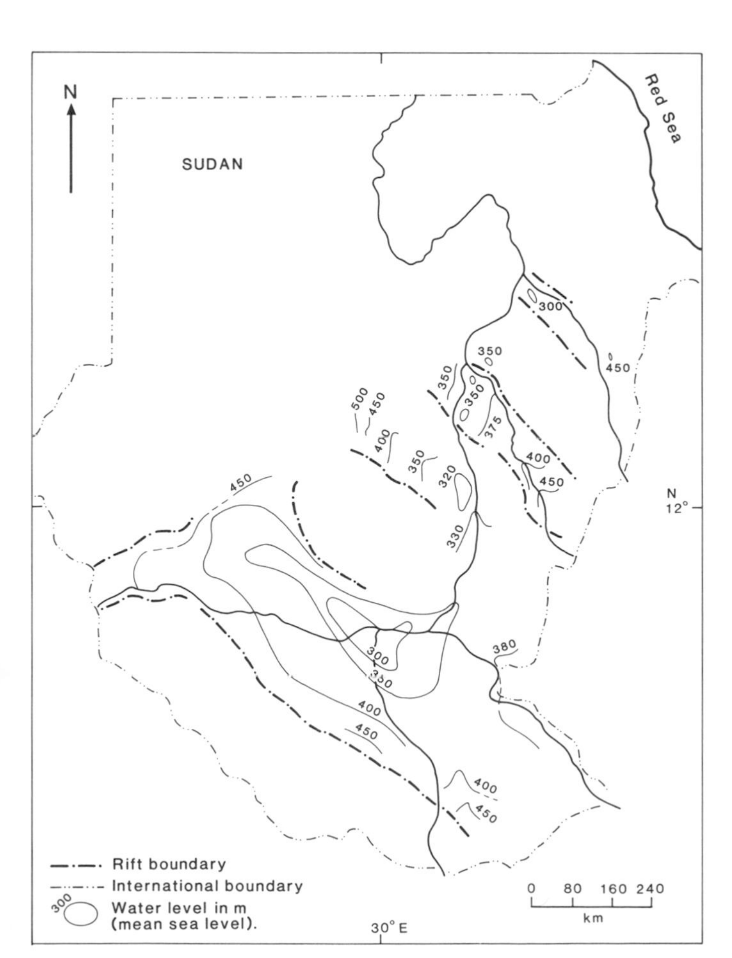

Figure 3. Groundwater troughs in the River Nile basins (modified from Salama, 1987).

At the flowing end of each of the rivers discharging in the rift system, alluvial fans were formed (Fig. 4): Gash and Atbara fans of the River Atbara and River Gash of the River Atbara rift; Soba and Gezira fans of the Blue Nile River in the Blue Nile rift; Abu Habil and Mashar fans of Abu Habil and Wadi Adar Rivers in the White Nile rift; El Kuu, El Ghalla, Shallengo, Nyala, Bulbul, and El Sudd fans of the Bahr El Arab River in Bahr El Arab rift. The deltas and fans were always found at the distal end of the graben or trough, against an uplifted block of basement, which seems to act as a sill or dam, forming a hydrologically closed basin, and causing the deposition of the sediments and the formation of evaporites and/or the saline lakes (Salama, 1985b, 1987).

Chemical controls

The study of the groundwater salinity in the rift systems shows that a saline groundwater body or bodies exist within the Tertiary sediments, these saline zones coincide with the deltas and fans (Fig. 5). The saline water bodies are identified as: (a) Sudd saline zone in Bahr El Arab rift; (b) Nuba and Adar saline zones in the White Nile rift; (c) Gezira and Soba saline zones in the Blue Nile rift; (d) Atbara and El Gash saline zones in the River Atbara rift.

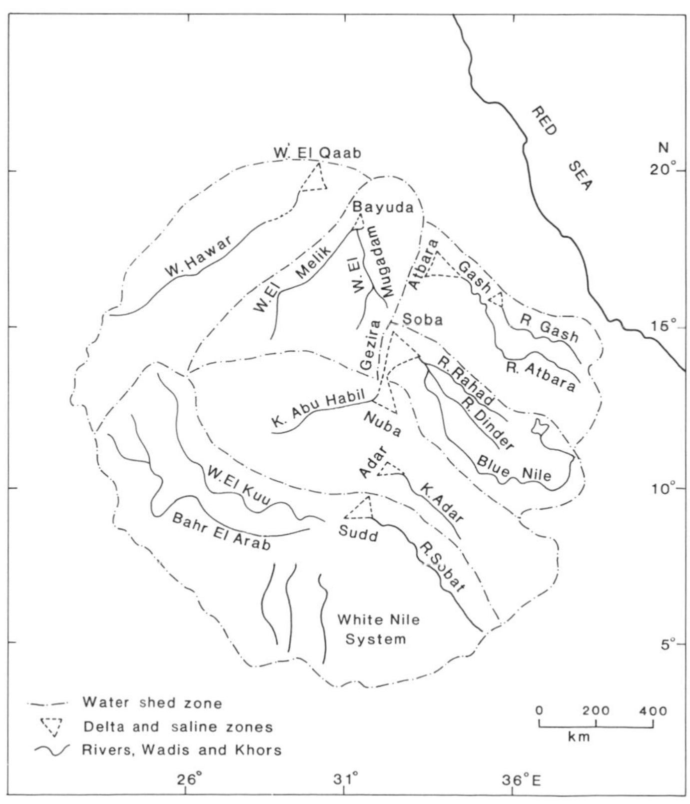

Figure 4. The Tertiary River Nile basins before the joining of the river system, showing the closed lake basins and river deltas in each rift structure. (K = Khor, R = River, W = Wadi)

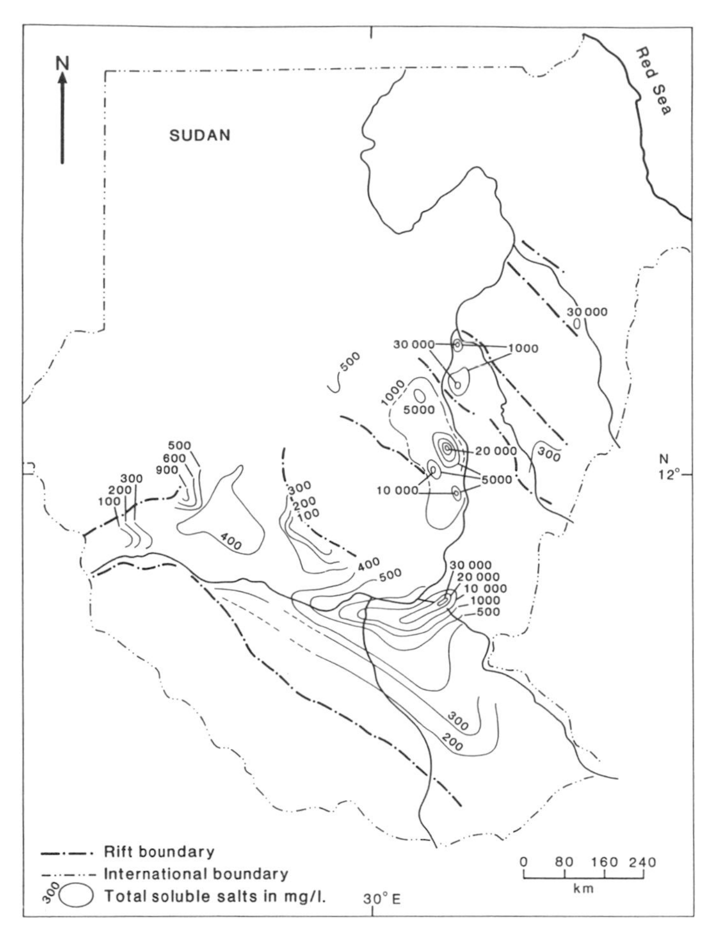

Figure 5. Groundwater saline zones in the River Nile basins (modified from Salama, 1987).

The Sudanese saline lakes were different from the present East Africa lakes. The East Africa rift saline lakes still exist, because they are in the high uplifted areas from which sediments are eroded and transported. The Sudanese saline lakes on the other hand, were in the lowest part of another subsiding rift system that collected vast amounts of sediments. The rapid rate of sedimentation accelerated the rate of burial of the saline lakes under a thick cover of Tertiary sediments. On the other hand, the rapid rates of deposition did not allow the lakes to become highly saline. During dry arid periods, the lakes were completely or partly evaporated to dryness, thus creating layers of high salinity, which were later dissolved by groundwater to form saline groundwater bodies. The alternating dry and wet periods caused the layering of the formations into saline and fresh zones, the wet periods were characterized by high rates of deposition that partially filled up the basins, and formed layers of fresh groundwater zones within and above the saline water bodies.

Saline lakes are very common phenomena in the East African rift system. Talling and Talling (1965), in reviewing the chemistry of these lake waters, have found that most of them are predominantly sodium carbonate–bicarbonate brines, ranging from dilute to highly concentrated. Magadi contains the most concentrated brines and thus represents an end stage in the development of Rift Valley Lakes (Eugster, 1970; Jones et al., 1977). Yuretich and Cerling (1983) in their study of the hydrochemistry of Lake Turkana, have shown that the high rates of sedimentation (up to 1 cm per year) may minimize the effects of diffusion between the interstitial waters and the lake water, and may cause significant removal of some of the cations and anions of the water. Cerling (1979) in his study of the palaeochemistry of Lake Turkana has shown that modern East African lakes follow a simple evaporation trend. Gac et al. (1977) in their work on evaporation from Lake Chad showed that most of the evaporating waters yield high concentrations of HCO_3-CO_3 and Na, but very low concentrations of Ca and Mg. An experimental evaporation of the diluted original water has shown that Ca depletion was due to calcite precipitation and Mg depletion was due to Mg-silicate formation. Other studies of the geochemical evolution of closed basin waters (Jones, 1966; Garrels and Mackenzie, 1967; Hardie and Eugster, 1970) have indicated that the solute composition of such waters is controlled primarily by silicate hydrolysis and evaporative concentration.

On the other hand, the buried saline lakes of Sudan have been formed through a continuous process of evaporation coupled with alkaline earth precipitation and resolution of capillary salt. The widespread presence of calcrete, kankar, and other carbonate deposits, over, and in, most of the Tertiary deposits, showed that conditions were favorable for the deposition of carbonates. It is postulated that the shallow standing waters evaporated, forming salt crusts. In the next flood, fresh water dissolved the most soluble salts (NaCl and Na_2SO_4) and transported these towards the deepest part of the basin, leaving carbonates behind in the form of kankar nodules. This explains the wide distribution of kankar nodules over the upper Tertiary horizons and the concentration of the sodium, chloride, and the sulphates in the saline zones.

At the same time, in the deepest part of the graben (which was always a lake, playa, or sebkha) the lake water evaporated, and became increasingly saline. During dry arid periods the lakes were completely or partly evaporated, thus creating layers of salts, which were later dissolved by groundwater to form saline groundwater bodies. Alternating dry and wet periods are well recorded (Kendall, 1969; Gasse, 1977; Gasse et al., 1980; Wendorf and Shild, 1976; Adamson et al., 1980; Livingstone, 1980). The wet periods are characterized by high rates of deposition, which partially fill up the basins and form zones of fresh groundwater within the saline water bodies. Due to the rapid sedimentation rates, the saline zones were quickly covered by sediments. These cycles eventually filled up the basins and led to their interconnection to form the existing Nile system.

Salinity is generally low at basin boundaries. In Bahr El Arab rift values less than 80 mg/L are recorded. Salinity increases gradually as the water moves into aquifer where it reaches values ranging between 500 to 800 mg/L some 500 km away from the boundary. This represents an increase of 1 to 2 mg/km of linear flow in the aquifer. This gradual increase is abruptly changed to sharp increases as the saline zones are approached (Salama, 1985a, b; Fig. 6). The relationship between chloride and sodium concentration of the Sudanese saline lakes has shown that simple evaporative conditions are controlling the concentration in the areas upstream of the saline zone (Salama, 1985b).

Salama (1985b) studied the salinity problem in the White Nile rift basin where water samples were collected from Kiteir Balla well at the depths 33, 46, 55, 61, 76, and 91 m. The results show that salinity increased with depth (Fig. 7). Three distinct zones were recognized: a top layer mainly recharged from the White Nile, a second layer formed of White Nile recharge water and brines, and the third layer of uncontaminated brines.

Thermodynamic saturation indices (SI: defined as the measure of departure from equilibrium SI = Log 10 [IAP/K] where IAP is the ionic activity product and K is the equilibrium constant) of the carbonate minerals, calcite ($CaCO_3$), dolomite ($CaMg[CO_3]_2$) and huntite ($CaMg_3[CO_3]_4$) and gypsum ($CaSO_4.2H_2O$) were calculated using WATSPEC (Wigley, 1977). It was found that all the samples were saturated with respect to calcite and dolomite and all the samples were unsaturated with respect to gypsum. It was found that where the effect of recharge is noticed, two samples were undersaturated with respect to huntite, whereas samples 5 and 6 were found to be saturated with respect to huntite. This suggested the possibility of using the metastable mineral huntite as a marker mineral for the determination of the brine areas, and to delineate the boundaries of the saline zones.

Huntite saturation from the saline zones of Soba, Gezira, Nuba, and Sudd were studied with respect to the isosalinity

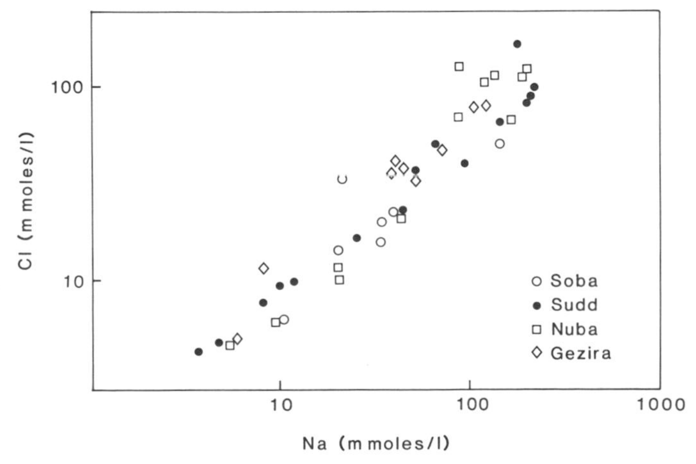

Figure 6. Na-Cl plot showing the gradual increase in salinity upstream of the saline zones and the abrupt increase at the saline zones.

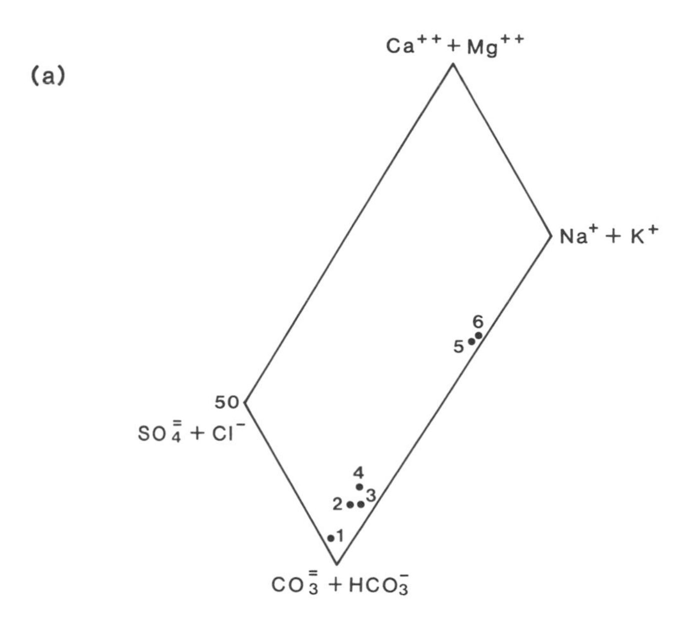

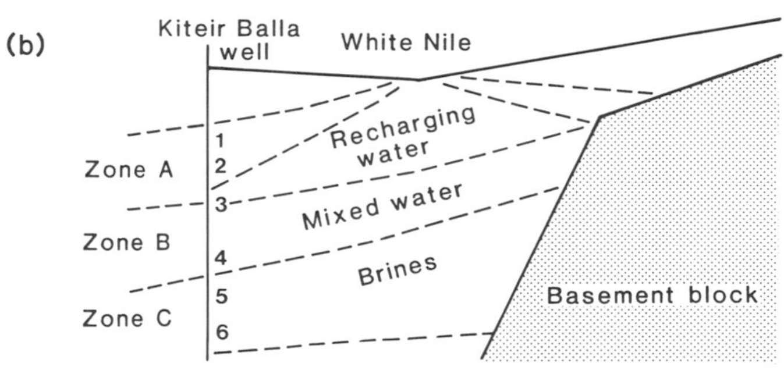

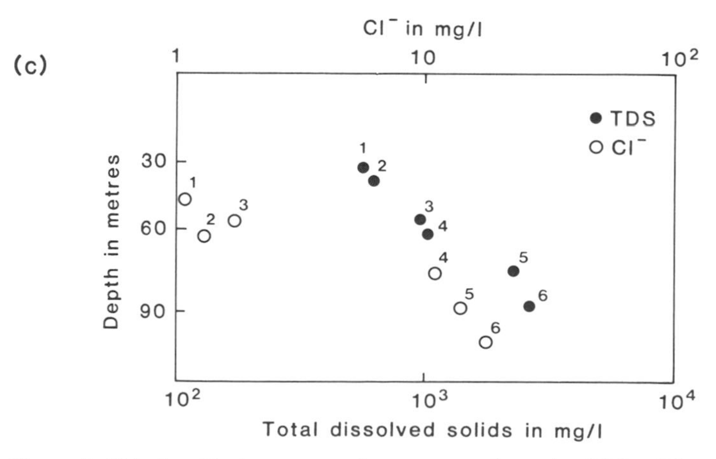

Figure 7. Relationship between recharge water from the White Nile and the brines in the Kiteir Balla area. Diamond shape plot (a) shows the brines from samples 5 and 6 plotting in a separate group. Location of sampling wells in relation to the basement block and the three zones of recharging water, mixed water and brine are shown in (b). The increase in salinity by depth is shown in (c).

map. It was found that isosalinity line of 1,000 mg/L, marks the huntite saturation point, which indicates that the SI of huntite together with the isosalinity line of 1,000 mg/L can be used as a marker to define the saline zone areas.

The increase in the sulphate content can be explained partly by the study carried out by Talling (1957) on the longitudinal succession of the physical and chemical water characteristics in the White Nile, especially in the Sudd region. He concluded that swamp effects included a partial deoxygenation and depletion of phosphate and sulphate. He further noted a significant removal of sulphate soon after the river water had entered the swamp region. Whether gypsum will be deposited in a lake depends on the concentration of sulphate and the concentration of Ca. The deposition of gypsum as reported in the Nuba and Sudd basins and the presence of high sulphate indicates that the saline lakes were saturated with respect to gypsum.

METHODOLOGY

The Sudanese lakes have had a complex history and undergone enormous changes in their size, sedimentation cycle, and environment of deposition through the different geologic periods. This study is based on the analysis of soil and water samples from about 900 exploration wells drilled on the top 500 to 1,000 m for rural water supply, from geophysical work carried by the geophysics section of the Department of Mines and their contracted companies. It covers the history of the saline lakes which developed during the post-rifting Sage phase.

GEOLOGY OF THE SUDANESE LAKES

Sudd lake basin

Basin configuration. Bahr El Arab rift comprises two major structures, the Baggara graben and the Sudd graben (Fig. 8; Salama, 1985a, b). Baggara graben covers the northern part of the rift and extends in an east-west direction. While the Sudd graben covers the southern part and extends in a northwest-southeast direction. The grabens and horsts indicate a steplike subsidence of separate blocks. The intensity of the faulting and the subsidence increase southward, where it attains a depth of more than 5 km in the Baggara graben and an estimated 11 km in the Sudd graben.

As outlined in Figure 8, the Bahr El Arab rift basin is an

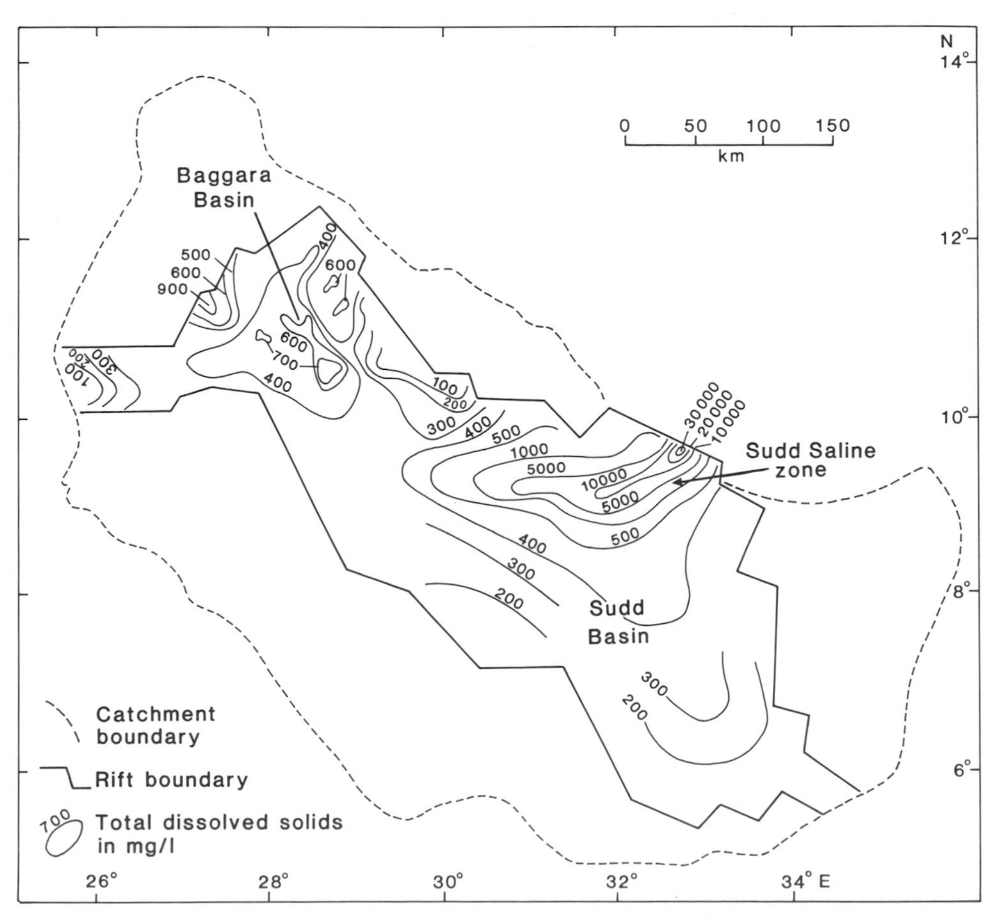

Figure 8. Bahr El Arab rift, showing Baggara basin in the north, Sudd basin in the south, and the saline Sudd zone at the Sudd swampy region.

elongate basin extending in a northwest-southeast direction. It is nearly 1,150 km in length, 600 km in width in its northern part and about 330 km in width in its southern part. It is surrounded by the high uplifted blocks: the Darfur dome in the north, the Nuba Mountains in the east, the Congo-Nile divide in the south and west, and the J. Marra massif in the northwest. The high lands would be similar to the high lands outlined by Holmes (1965). Faure (1975) calculated the rate of uplift of the East African highs to be about 0.1 mm/yr in the mid-Tertiary. Using this figure to calculate rates of uplifts in the Sudanese blocks gives the following heights: 2,500 m for the Congo-Nile uplift, 3,800 m for the Nuba Mountains, 1,500 m for the Darfur dome, and 2,000 m for J. Marra massive. The highs were eroded depositing sediments in Bahr El Arab troughs.

Sediments. Williams and Williams (1980) in their synthesis of Nile evolution showed that extensive alluvial fans were built up in southern and central Sudan at the outlet of major river issuing from the highstands of Ethiopia. Adamson and Williams (1980) showed that the surface of Bahr El Arab rift is covered by active and abandoned alluvial fans, by swamp deposits, by active distributary streams, and by prior stream channels. They described a series of low-angle alluvial fans and trough sediments in the boundary of the Nuba Mountains and the Babanusa trough.

The outcrops and top cover of the Tertiary sediments are formed of unconsolidated sands, gravels, silts, and clays deposited in alluvial, fluvial, and shallow lacustrine environments (Vail, 1978). The sediments were classified (Salama, 1985a) into distal-fan deposits, alluvial deposits, and lacustrine deposits. The distal-fan deposits were deposited at the flowing end of all major rivers and wadis. These are formed of clays, with small subangular to subrounded basement fragments, small iron concretions, volcanic chips, and kankar nodules. The alluvial deposits are formed of well-rounded gravel, pebbles, and conglomerates, interbedded with sand and clay. The lacustrine deposits are formed of dark green gravely clays with veneers of whitish sand; very fine silt and intercalated layers of sand, clay, and silt; and layers of gypsum and evaporites. Thick carbonate deposits were recorded (Salama, 1985a) from shallow wells and drilled boreholes in the western areas of the basin. The chemical analysis of the carbonate sample showed that it is mainly calcium carbonate with small percentages of chlorides and sulphate (Table 1). Calcite was found to be the main carbonate mineral.

The subsurface Quaternary and Tertiary sediments are divided into three groups (Anon, 1981a, b, 1982; Salama, 1985b; Schull, 1988; Fig. 9):

a. Recent–upper Miocene sediments range in thickness from 600 to 1,000 m of poorly sorted, iron-stained sands and siltstone with interbedded clays, changing southward to light colored (volcanic ash) claystone. This active period of deposition was followed by a hiatus, and a period of slow deposition.

b. Oligocene–upper Eocene sediments range in thickness from 600 to 3,000 m of interbedded, clear quartz sands and multicoloured clays. Local development of lacustrine shale in axial regions of rift basin.

c. Eocene-Paleocene sediments range in thickness from 600 to 3,000 m, of massive sand deposits composed mainly of coarse clear and white quartz grains with locally abundant pyrite.

TABLE 1. CHEMICAL ANALYSIS OF A CARBONATE SAMPLE FROM THICK CARBONATE DEPOSITS RECORDED IN SHALLOW WELLS AND DRILLED BOREHOLES FROM THE WESTERN AREAS OF THE SUDD LAKE BASIN

Cations (mg/kg)			
Ca	Mg	Na	K
122,000	4,680	4,680	2,810
Anions (mg/kg)			
CO_3	Cl	SO_4	
289,000	925	2,033	

Sudd lake. Salama (1987) showed that a series of freshwater lakes exists at the edges of Bahr El Arab; including Lake Keilak, Lake Abyad, and Lake Kundi, together with the main Sudd Lake, which covers most of the central part of this river system. Based on hydrological data he concluded that if the Victoria Nile was not connected to the White Nile the Sudd would be a closed-lake system.

Independent evidence showing that the White Nile was not connected to the main Nile prior to 12,500 yr B.P. was presented by Shukri (1949) based on mineralogical analysis of the Nile deposits and by Kendall (1969), Livingstone (1980), and Adamson and Williams (1980) based on hydrological and geological evidence. This evidence indicates that the Sudd depression of southern Sudan was a closed-lake system.

Willcocks (1904) postulated that the ancient lake was 402 km (250 m) in length from north to south and that the Blue Nile flowed southwards to join this lake. Lawson (1927) elaborated the Lake Sudd hypothesis and was first to call it by this name. Ball (1939) developed the Lake Sudd hypothesis further, assigned a length of over 1,054 km (655 m) to the lake. Ball (1939) made calculations similar to those of Lawson (1927) and concluded that for a lake of these dimensions, an average evaporation rate of 3 mm over the lake would be sufficient to dispose of all rain and river water entering the lake, because the average annual rate of evaporation from open water surfaces at Mongalla, Malakal, and Khartoum are 3.0, 4.5, and 7.5 mm, respectively (Hurst and Phillips, 1931).

Berry and Whiteman (1968) and Whiteman (1971) went to great lengths to prove that Lawson and Ball were wrong. Salama (1987) agreed that the lake may never have achieved the size assumed by the pioneering workers, yet there is enough evidence to show that there was always a water body in the Sudd region (Berry, 1962; Salama, 1987). However, the lake was not formed by a dammed up Nile. It was formed in a closed basin caused by subsidence in Bahr El Arab rift during

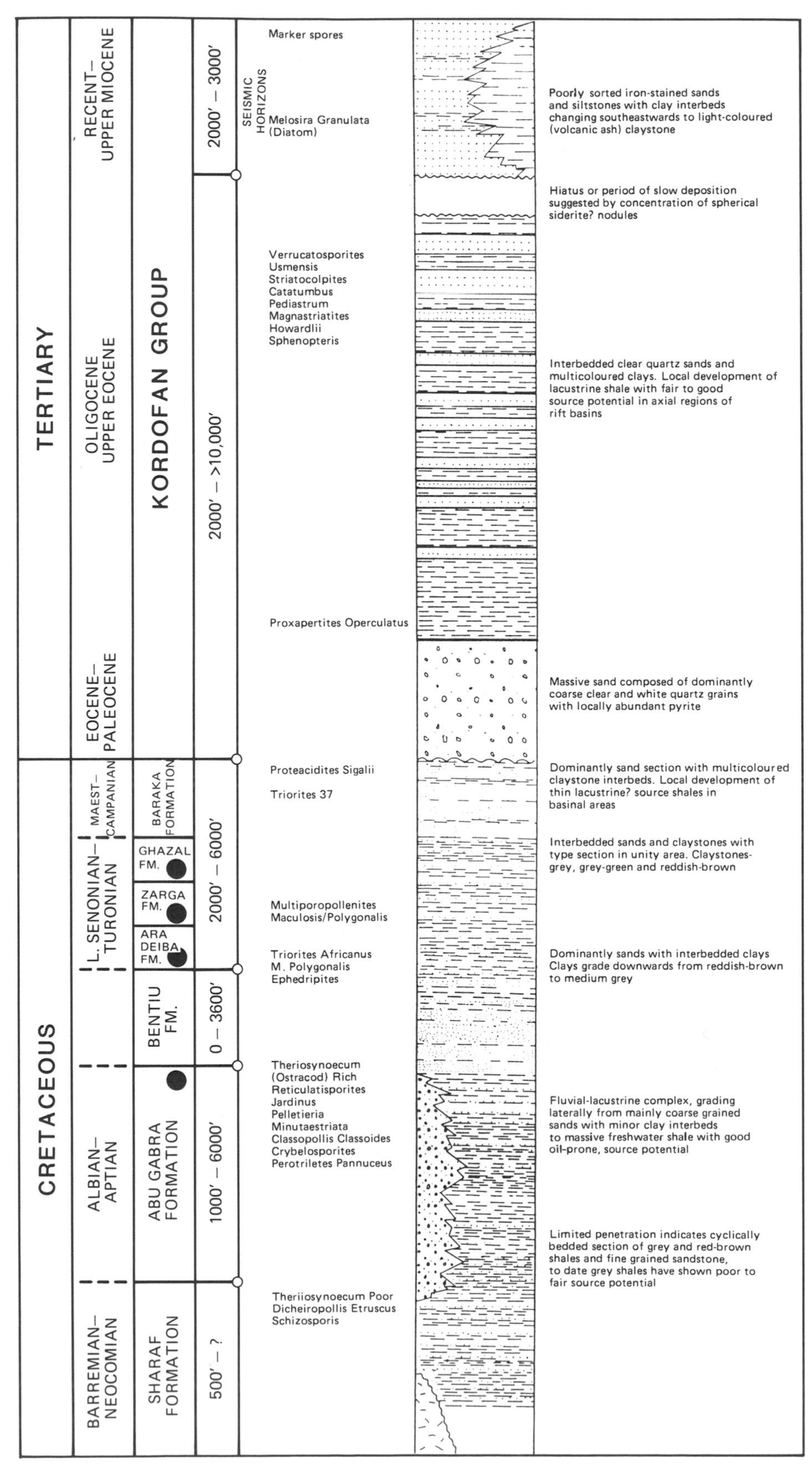

Figure 9. Typical section in Bahr El Arab rift.

the Sage period (Salama, 1985a, b). The size fluctuated according to the palaeoclimatological events prevailing at the time.

Salama (1987), using saturation indices of minerals and salinity parameters (Fig. 10), showed that a saline lake occupied the central area of the Sudd. He estimated that the lake fluctuated in size from a maximum of 32,000 km^2 to a minimum of 336 km^2 (the area enclosed between isosalinity lines of 1,000 and 30,000 mg/L). The depth of the lake ranged from few meters at the borders of the lakes to about 200 m in the central part (calculated from groundwater gradients, versus thickness of saline zone). The presence of thick carbonate deposits at Bahr El Arab, suggests that the size of the lake was greater than the figures calculated or perhaps indicates the presence of another separate saline lake in that part.

Salama (1987) reported the presence of carbonate deposits 0.5 to 1.5 m thick, along the banks of Bahr El Arab. Thick lacustrine carbonate deposits at the western edge of Bahr El Arab rift, where a known fault zone was also recorded from shallow wells and borehole records (Salama, 1987). The presence of a closed saline lake might be responsible for the deposition of the carbonate by contact of calcium-rich fresh water with saline water bodies (Sly, 1978). It is possible that, as lakes always form on the down-faulted part of the fault zone that is continuously subsiding, the surface drainage entering the lake in the form of sheet flow will deposit the carbonate at the contact zone between the fresh and saline water bodies. This might also explain the concentration of sulphate in the central part of the lakes. As the water will be depleted in its carbonate content, the central part will continuously exhibit increase in sulphate and chlorides. The widespread occurrence at various depths of carbonate indicate that this area was always a contact zone between the saline areas and the fresh waters entering the lakes.

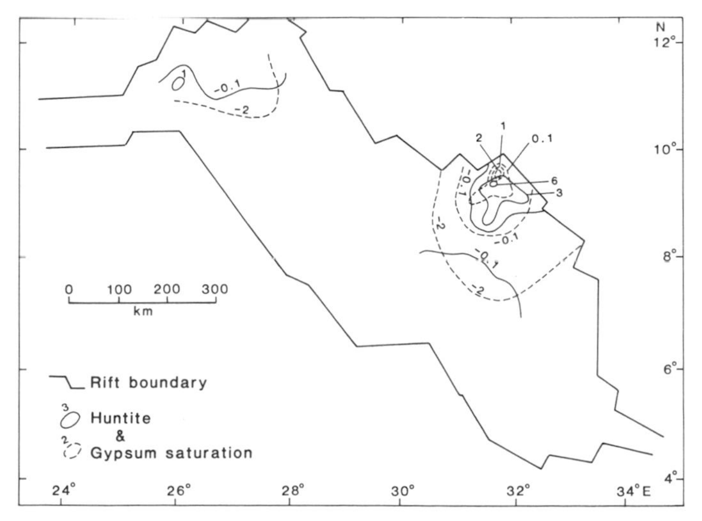

Figure 10. Saturation indices (see text for explanation) of huntite and gypsum for the Sudd basin used to delineate the boundaries of the saline lake. Saturation indices (S.I.) are dimensionless numbers.

Nuba lakes basin

Basin configuration. The White Nile rift is formed by the junction of two major grabens; the Umm Ruwaba graben extending in a northwest direction and the White Nile graben extending in a north to northwest direction (Fig. 11). It is bounded in the north by east-west faults and northeast-southwest fault systems. At Bara, farther south, another fault system is known from geophysics with a northerly downthrow of 600 m (Ali, 1978). An uplifted basement block extending in a northwest-southeast direction divides the graben into two troughs; the Bara trough in the east and Umm Ruwaba Renk trough in the southeast. The Umm Ruwaba trough is characterized by a series of horsts and grabens formed by two sets of fault systems that trend east-west and northwest-southeast (REGWA, 1979). The basement rocks and the overlying Tertiary deposits are block faulted upwards at Rabak, whereas the Tertiary sediments are downthrown more than 100 m at Kosti. Grabens and horsts bounded by northwest-southeast fault systems form a series of troughs of the White Nile graben (Geophysics and Strojoexport, 1967, 1977; Anon, 1981a, b, 1982).

Sediments. The White Nile rift basin is elongate in shape and extends in a northwest-southeast direction. It is about 520 km in length and 400 km in width at Umm Ruwaba trough and 360 km in width in the White Nile trough (Fig. 11). The fine deltaic deposits of Khor Abu Habil were deposited in the Nuba delta (Fig. 4). Coarse sediments were deposited in the deeper parts of the troughs, and are encountered at various depth intervals from the surface to 600 m (Salama, 1985a, b).

The sediments deposited in Khor Abu Habil are characterized by the absence of volcanic clasts and the high percentage of coarse material in the margin of the fault zones at a depth of 200 to 300 m, the high percentage of fines and clays in the fan area, and the presence of salt layers (evaporites) in the fan delta area. There are several published descriptions of the sediment types in the Umm Ruwaba graben and these appear to vary with the well location. Most workers suggested that the Umm Ruwaba sediments are fluviatile and lacustrine (Andrew, 1948; Rodis et al., 1963; Salama and Salama, 1974). Whiteman (1971) suggested that the sediments were laid down in a series of land deltas. Salama (1985a) showed that the Umm Ruwaba sediments have multiple origins with layers of very fine sand (possibly aeolian), and layers of gypsum and evaporites reported in several boreholes.

Marsail Salama (personal communication, 1985) in a detailed interpretation of Landsat imagery for Khor Abu Habil showed the presence of palaeochannels extending from Umm Ruwaba eastwards with main tributaries towards the northeast to Kosti and southeast to Keri Kera. She also showed that Abu Habil fan deposits are mainly clays and very fine silts. Similar deposits are known in most of the internal drainage systems of today's major wadis, such as the River Gash, Wadi El Kuu, Wadi Nyala, Wadi Bulbul, and Wadi Shellango (Salama, 1985b, 1987).

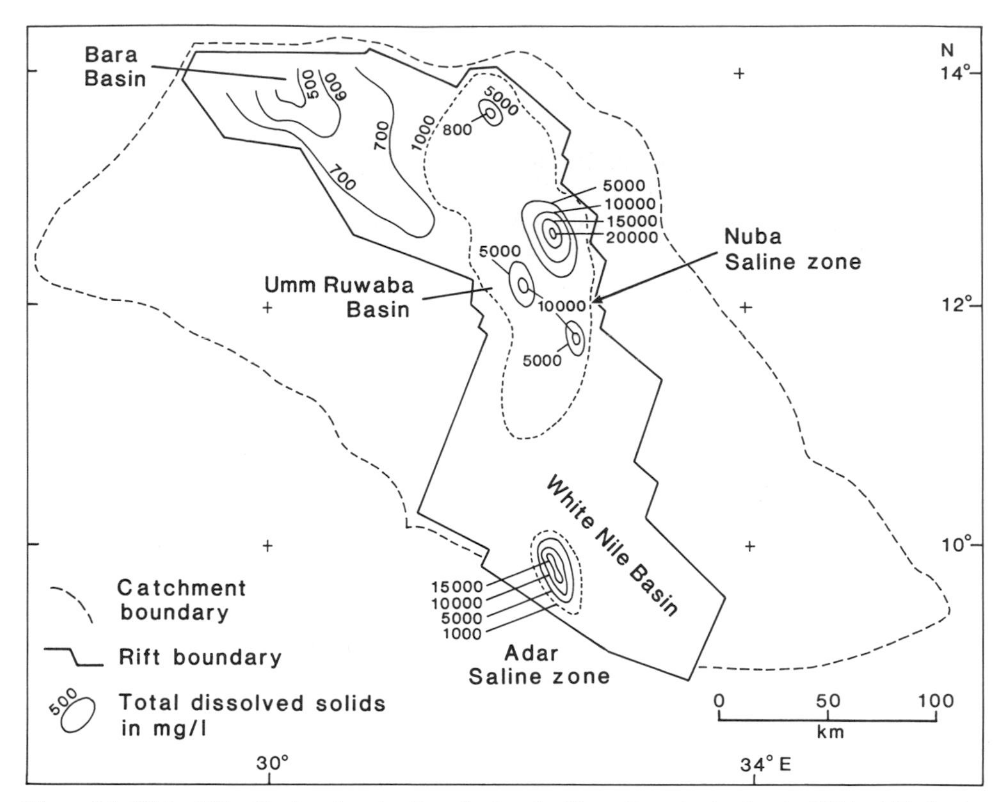

Figure 11. White Nile rift showing the Bara basin, the Umm Ruwaba basin, the White Nile basin, and the Nuba saline zone.

Thick extensive carbonate deposits are reported at Khartoum (Williams and Anderson, 1980), White Nile near Dueim (Salama, 1985b), and Gezira (Adamson et al., 1982; El Boushi and Abdel Salam, 1982).

Nuba lakes. The Nuba buried saline lakes extend from near Abu Habil in the north, to Wadi Adar in the south (Fig. 4). They were formed by the palaeo-Abu Habil and Wadi Adar drainage system. Today the mean annual flow of Abu Habil is 100 million m^3, while that of wadi Adar is estimated to be 1.5 to 2.0×10^9 m^3 (National Council for Research, 1982).

Landsat satellite imagery (October, 1972) shows a large alluvial fan on the left bank between Kosti and Keri Kera with numerous small meandering distributaries that have been formed by Khor Abu Habil (Gunn, 1982). Salama (1985b) showed that the trough which occupies the central part of the White Nile Rift is the lowest water level contour (320 m) in the central part of Sudan. This indicates that the Nuba lakes received water from the Gezira and Sudd lakes at one time or another during its depositional history. This also seems likely because the very slow deposition rates of Abu Habil compared to the White and Blue Niles system always kept the Nuba lakes area at a lower depositional level.

Salama (1985b) using salinity parameters and saturation indices (Figs. 5 and 12) postulated that the area enclosed by the isosalinity line of 1,000 mg/L defines the margins of the saline lake zone. The high salinity zones within this saline zone represent four smaller separate buried saline lakes. Each of these lakes occupied one of the troughs within the rift structures of the Umm Ruwaba graben (Abu Habil saline zone in Abu Habil trough, East and West Er Rawat saline zones in East and West Er Rawat troughs). Another saline water body along Wadi Adar, in the Adar trough, with salinities higher than 10,000 mg/L, indicates the presence of another saline lake in this trough. The presence of high sulphate concentrations, in the Nuba lakes show that they had a similar depositional environment to that at the Sudd Lake.

Williams and Adamson (1980) in their study of the late Pleistocene evaporites near Esh Shawal indicated that the absence of clastic components, the well-ordered crystalline structure and uniform altitude of the evaporites, and their radiocarbon ages all suggest prolonged late Pleistocene evaporation of an extensive body of still, saline water along what is now the White Nile flood plain. The rough accordance in age between these White Nile evaporites and those that crop out in the Blue Nile south of Khartoum may indicate prolonged aridity and an absence of outflowing drainage in the Gezira at this time. The radiocarbon ages are more than 40,000 years (Adamson et al., 1982).

Blue Nile lakes basin

Basin configuration. The Blue Nile rift (Fig. 13) extends from south of the Sabaloka gorge "The Sixth Cataract" south-

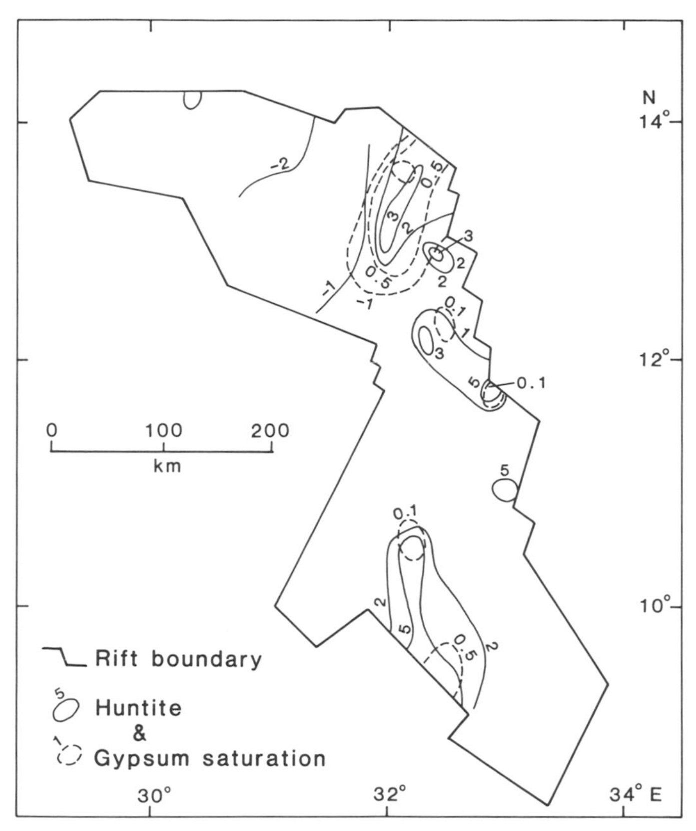

Figure 12. Saturation indices of huntite and gypsum for the Nuba basin used to delineate the Nuba saline zones.

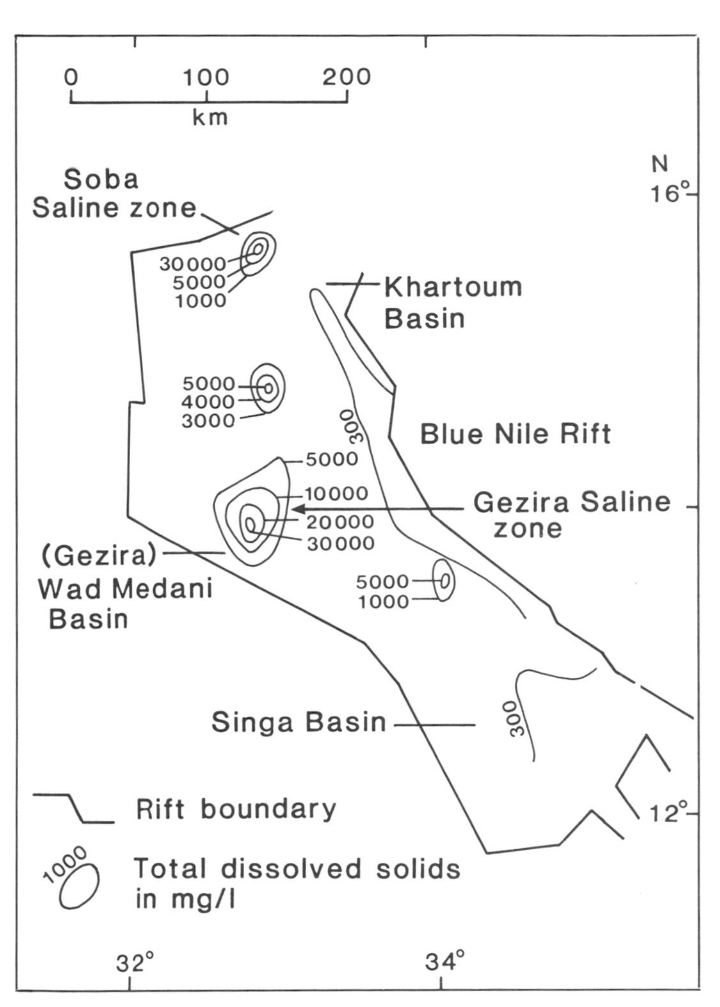

Figure 13. Blue Nile rift showing Khartoum, the Wad Medani (Gezira) and Singa basins, and Soba and Gezira saline zones.

east to the Sudan borders with Ethiopia, following the Blue Nile, River Rahad, and River Dinder. It is formed of three main grabens, from north to south, Khartoum, Wad Medani, and Singa, with depth increasing southward. Its southern limit is the elevated Basement block extending in a northwest-southeast direction with various hills cropping out along this boundary line. The northern margin is another elevated block of basement rocks extending in a northwest-southeast direction east of River Rahad. A series of fault systems striking northwest/southeast parallel to the Blue Nile, River Er Rahad, and River Dinder having westerly downthrows with southerly increase in depth.

Sediments. The Blue Nile rift basin is an elongate basin extending in a northwest-southeast direction, it is about 350 km in length and ranges in width from 80 km in the northern part to 180 km in the southern part. The Blue Nile drains the central and northern part of the high Abyssinian plateau. It is joined in Sudan by the Rahad and Dinder tributaries. The Blue Nile and its tributaries are characterized by a very high suspended sediment load during the flood season of 4,000 mg/L in August, and 100 mg/L during the recession period (El Badri, 1972). The sediments of the Blue Nile are estimated to be 41 million tonnes annually (Hurst and Phillips, 1931). The sediments filled the Blue Nile rift during the Tertiary and possibly part of the Quaternary epochs. As concluded from the previous section, following the cessation of subsidence and infilling of the grabens and troughs of the Blue Nile rift, sedimentation proceeded to move southward towards the White Nile rift, to fill the Abu Habil trough.

The sediments in the Blue Nile rift have been studied by different workers (Andrew, 1948; Kheiralla, 1966; Williams, 1966; Whiteman, 1971; Williams and Adamson, 1973, 1980; El Boushi and Abdel Salam, 1982; Adamson et al., 1982; Salama, 1985b).

Williams and Adamson (1980, p. 290) provide the following description. "With few exceptions, the late Pleistocene and Holocene sediments bordering the present Blue Nile between Sennar and Khartoum represent an upward-fining fluviatile sequence. The Holocene alluvium consists of dark, alkaline cracking clays rich in subfossil shell fragments; and the late Pleistocene sediments generally comprise current-bedded fine, medium and coarse sand, with extensive outcrops of massive calcium carbonate in bank sections between Hasaheisa and

Khartoum, and more localized pockets of water transported volcanic ash". Abdel Salam (1966) recognized three subdivisions that were not consistent with lithological logs of more than 500 wells, studied in the Gezira area (Salama, 1985b). The deposits were formed of layers of clay, sand clay, clayey sand, sand, and gravel, a feature that is characteristic of alluvial fans and deltas that cannot be separated into divisions. The clays are alkaline, dark in color, and low in organic matter (Ruxton, 1965). They contain a high proportion of unweathered minerals such as calcic plagioclase, titan-augite, hornblende, and brown biotite (Andrew, 1948). The sand is composed of angular to subangular quartz grains, with calcium carbonate nodules and mica. The gravels are predominantly quartz along with fragments of metamorphic and igneous rocks, basalts, and agates as well as kankar nodules. Mcdougall et al. (1975) estimated the volume of the Gezira fan to be about 1,800 km^3. They also postulated that the total volume of material deposited by the Nile and Atbara in Egypt and Sudan is between 98,000 and 225,000 km^3, the bulk of which form the Nile Delta. They concluded that the close agreement between the volume of material eroded from Ethiopia and the volume in the Nile Delta is further evidence suggesting that the bulk of the Nile Delta sediments are of Ethiopian origin.

Recent exploratory drilling for oil in the Dinder area has revealed that the Tertiary sediments can be more than 3 km thick. The volume of sediments in the Dinder trough, and those required to fill up the White Nile troughs, would easily account for the volume of sediments estimated by Mcdougall et al. (1975) to have been deposited in the Nile delta. This would change completely the hypothesis put forward by the authors that the sediments in the Nile delta are of Abyssinian origin only. As a matter of fact, the new evidence suggests completely different sources.

Williams and Adamson (1980) described the Gezira (the area between the Blue and White Nile) as a complex low-angle alluvial fan of a type common in other semiarid areas of the world. They recognized three upward-fining alluvial cycles. Adamson et al. (1982) showed the presence of palaeochannels of the Blue Nile, Rahad, and Dinder. These can be seen in Figure 14. The palaeochannels of the Blue Nile show two branches very clearly. One is parallel to the recent Blue Nile and terminates in a fan form in the Soba area, while the other moves west toward the recent White Nile and some of its branches move south west, towards Ed Dueim.

Gezira lake. Salama (1985b) showed that the area west of Ed Dueim is characterized by the presence of recent sand dunes; below those sand dunes thick layers of evaporites are present. Results of analysis of samples collected from these evaporites are shown in Table 2. The analysis shows that the deposits are very similar to the carbonate deposits near Bahr El Arab (Salama, 1987). El Boushi and Abdel Salam (1982) noted that the presence of gypsum and carbonates in the Gezira sediments imply that saline waters accumulated in an internal drainage basin in the past.

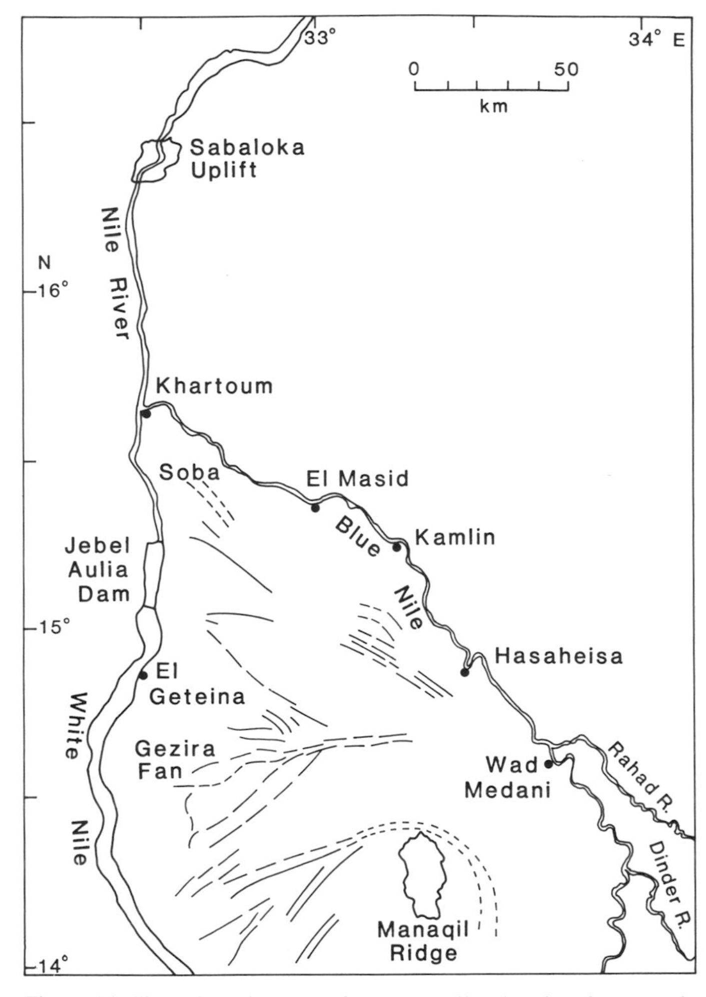

Figure 14. The palaeochannels of the Blue Nile, showing the westerly and southwesterly trend of the river towards the Gezira saline zone.

TABLE 2. ANALYSIS OF EVAPORITE DEPOSITS FROM WEST OF DUEIM

		Cations (mg/kg)	
Ca	Mg	Na	K
256,000	5,270	2,470	2,440
		Anions (mg/kg)	
CO_3	Cl	SO_4	
179,000	1,456	4,487	

Adamson et al. (1982) reported the presence of thick carbonate deposits in the Gezira; they also showed that the conditions were favorable for the deposition of carbonates during much of the Pleistocene and probably earlier. The presence of massive carbonate deposits or calcrete suggest prolonged and high input of dissolved carbonate under conditions suitable for its precipitation. These were deposited by sluggish stream flow and wide dispersal of the water at the down-slope end of a fan such as the Gezira.

Salama (1985b) postulated that the high salinity zone which is located in Matug trough and roughly coincides with the left branch of the palaeochannel of the Blue Nile (Fig. 14) forms the Gezira buried saline lake. Using saturation indices (Fig. 15) he also postulated that the lake size would fluctuate between a minimum of 250 km^2 to a maximum of 15,000 km^2.

From the presented evidence, it is clear that the Gezira depression was a closed basin for a long time during the Pleistocene. The cyclic pattern of wet and dry periods during the late Tertiary and Quaternary is well recorded (Gasse, 1977; Gasse et al., 1980; and also noted above). Adamson (1982) has shown that these cyclic events are applicable to all east and central parts of Africa and can even be extended to the north part of Africa. This cyclic pattern is reflected in the Gezira area as well as in the other buried lakes in Sudan.

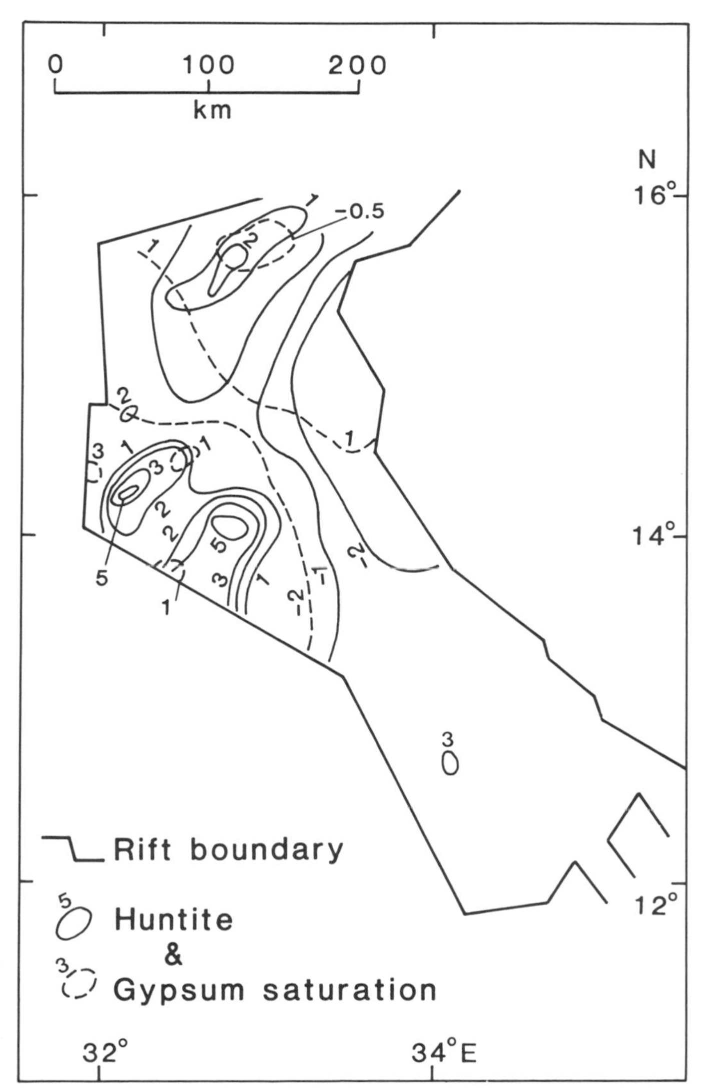

Figure 15. Saturation indices of huntite and gypsum used to delineate the Soba and Gezira saline zones.

Soba lake. Williams and Adamson (1980) described the extensive presence of thick late Pleistocene carbonate deposits cropping out in the west bank of the Blue Nile some 5 to 8 km upstream of Khartoum. They also described the presence of saline, alkaline, calcareous, sandy, and massive clays sediments in the area between Khartoum and El Masid. This saline zone also extends northward on both sides of the River Nile, the salinity increasing to very high levels near Sabaloka. Salama (1985b) showed that Soba Lake covers the area extending from south of Sabaloka cataract in the north to the margins of the Khartoum graben in the south. Based on soil and groundwater salinity he postulated that all this area was covered by evaporite deposits. During wet periods the main Nile eroded its course in the Sabaloka cataract, the surface water passing through the northern outlet caused the leaching of salts from the top layers. This leaching is more noticeable in the top 10 m of sediments 2 to 5 km east and west of the Blue Nile areas.

Relative age of the saline lakes and relation to East Africa lakes

Is it possible, in a preliminary way, to correlate the saline lakes in Sudan with similar lakes in eastern Africa, Ethiopia, and Afar.

a. All radiocarbon ages determined from groundwater in the central parts of these basins, gave an age of above 40,000 years B.P. (Salama, 1985b; Malmberg and Abdel Shafie, 1975).

b. Malmberg and Abdel Shafie (1975) using $\delta^{18}O$ results of groundwater from Bara basin showed that the groundwater from the lower horizon is from rains of much wetter and cooler periods than the present, with the temperature 4 to 8 °C cooler.

c. Assuming equal rates of erosion and sedimentation throughout the sedimentary column during the Quaternary, the top 100 to 300 m of sediments, which contain the saline zones, would accumulate in 10 to 30 ka using the rate of sedimentation of Yuretich and Cerling (1983), and 100 to 300 ka on an average rate of 1 mm/yr.

d. The approximate age of the lakes was calculated by the amount of salt deposited as related to the annual flow. Assuming the salt content of the inflow is 120 mg/L (which is comparatively high even in today's rates), and the palaeoflow would range from 2 to 20 km^3 annually, then the lake age would range from 40 to 110 ka.

e. From the rough estimates of ages made by sedimentation rates of item c above and the age determined by the salinity data d, the approximate age of the Sudanese lakes may be comparable with the ages of the lakes Abhe I and II (Gasse, 1977), which is between 30 and 70 ka. The Sudanese lakes can also be correlated with the wet pluvial periods of Said (1981). From this limited evidence of age, it can be postulated that the formation of those saline water bodies would most probably be in the period between 90 to 120 ka, that is, within the late Quaternary.

CONCLUSIONS

The study of the groundwater salinity in the rift basins has shown that they coincide with the deltas and fans which occupy the flowing end of the hydrological and hydrogeological closed basins. The deepest part of the basin at the downstream end was always a lake, sebkha, or playa. The saline deposits were formed through processes of efflorescence, leaching, and evaporation, leaving behind carbonate deposits as kankar nodules in the sheet flow area, and as thick deposits in the marginal fault zones.

ACKNOWLEDGMENTS

The author wishes to thank Professor Rushdi Said (Intergeosearch, USA) and Associate Professor D. Adamson (Macquarie University, Australia) for their constructive reviews, and Dr. M. R. Rosen (University of Minnesota) for editorial handling.

REFERENCES CITED

Abdel Salam, Y., 1966, The groundwater geology of the Gezira [M.Sc. thesis]: Khartoum, Sudan, University of Khartoum.

Adamson, D. A., 1982, The integrated Nile, *in* Williams, M.A.J., and Adamson, D. A., eds., A land between two Niles: Rotterdam, Balkema, p. 221–234.

Adamson, D. A., and Williams, F., 1980, Structural geology, tectonics and the control of drainage in the Nile Basin, *in* Williams, M.A.J., and Faure, H., eds., The Sahara and the Nile: Rotterdam, Balkema, p. 225–252.

Adamson, D. A., Gasse, F., Street, F. A., and Williams, M.A.J., 1980, Late Quaternary history of the Nile: Nature, v. 288, p. 50–55.

Adamson, D. A., Williams, M.A.J., and Gillespie, R., 1982, Palaeogeography of the Gezira and of the lower Blue and White Nile valleys, *in* Williams, M.A.J., and Adamson, D. A., eds., A land between two Niles: Rotterdam, Balkema, p. 221–234.

Ahmed, F., 1982, Implications of Precambrian lineaments on the Red Sea tectonics based on landsat study of northeast Sudan: Global Tectonics and Metallogeny, v. 1, p. 326–335.

Ali, H. O., 1978, Gravity exploration and numerical simulation of groundwater resources in Bara basin [M.App.Ssc. Project Report]: Kensington, N.S.W., Australia, University of New South Wales.

Almond, D. C., 1982, New ideas on the geological history of the basement complex of north-east Sudan: Sudan Notes and Records, v. 59, p. 106–136.

Andrew, G., 1948, The geology of the Sudan, *in* Tothill, J. D., Agriculture in the Sudan: London, Oxford University Press, p. 84–128.

Anon, 1981a, Local scoring a big Sudanese concession: Oil and Gas Journal, Feb. 2, 1981, p. 34.

Anon, 1981b, Chevron plans first field development: Oil and Gas Journal, Sept. 14, 1981, p. 46.

Anon, 1982, Search simmering in several areas of Sudan, Parker, M., ed.: Oil and Gas Journal, May 17, 1982, p. 36.

Ball, J., 1939, Contributions to the geography of Egypt. Cairo: Gizeha, Egyptian Government Press, 308 p.

Berry, L., 1962, Large scale alluvial islands in the White Nile: Revue Geomorphologie Dynamique, v. 12, p. 105–108.

Berry, L., and Whiteman, A. J., 1968, The Nile in the Sudan: Geographical Journal, v. 134, p. 1–33.

Browne, S. E., and Fairhead, J. D., 1983, Gravity study of the central African rift system: A model of continental disruption: 1, The Ngaoundere and Abu Gabra rift, *in* Morgan, P., ed., Processes of continental rifting: Tectonophysics, v. 94, p. 187–203.

Browne, S. E., Fairhead, J. D., and Mohamed, I. I., 1985, Gravity study of the White Nile rift, Sudan, and its regional tectonic setting: Tectonophysics, v. 113, p. 123–137.

Cerling, T. E., 1979, Palaeochemistry of Plio-Pleistocene Lake Turkana, Kenya: Palaeogeography, Palaeoclimatology, Palaeoecology, v. 27, p. 247–285.

El Badri, O., 1972, Sediment transport and deposition in the blue Nile at Khartoum, flood seasons 1967, 1968, 1969 [M.Sc. thesis]: Khartoum, Sudan, University of Khartoum.

El Boushi, I. M., and Abdel Salam, Y., 1982, Stratigraphy and ground water geology of the Gezira, *in* Williams, M.A.J., and Adamson, D., eds., A land between two Niles: Rotterdam, Balkema, p. 65–80.

Eugster, H. P., 1970, Chemistry and origin of the brines of Lake Magadi, Kenya: Mineralogical Society of America Special Paper, v. 3, p. 213–235.

Faure, H., 1975, Recent crustal movements along the Red Sea and Gulf of Aden coasts in Afar (Ethiopia and TFAI): Tectonophysics, v. 29, p. 479–486.

Gac, J. Y., Droubi, A., Fritz, B., and Tardy, Y., 1977, Geochemical behavior of silica and magnesium during the evaporation of waters in Chad: Chemical Geology, v. 19, p. 215–228.

Garrels, R. M., and Mackenzie, F. T., 1967, Origin of the chemical composition of some springs and lakes, *in* Gould, R. F., eds., Equilibrium concepts in natural water systems: American Chemical Society Advances in Chemistry Series, no. 67, p. 222–242.

Gasse, F., 1977, Evolution of Lake Abhe (Ethiopia and TFAI), from 70,000 B.P.: Nature, v. 265, p. 42–45.

Gasse, F., Rognon, P., and Street, F. A., 1980, Quaternary history of the Afar and Ethiopian rift lakes, *in* Williams, M.A.J., and Faure, H., eds., The Sahara and the Nile: Rotterdam, Balkema, p. 361–400.

Geophysics and Strojoexport, 1976, Geophysical investigations of groundwater structures, western part of Kordofan province and eastern part of Darfur province. 4 stages: Khartoum, Report submitted to Rural Water Corporation, (Unpublished), 107 p.

Geophysics and Strojexport, 1977, Geophysical investigations of groundwater structures in central and northern part of the Upper Nile Province—Fifth Stage: Khartoum, Report submitted to Rural Water Corporation. (Unpublished), 86 p.

Gunn, R. H., 1982, The plains of the central Sudan, *in* Williams, M.A.J., and Adamson, D. A., eds., A land between two Niles: Rotterdam, Balkema, p. 81–109.

Hardie, L. W., and Eugster, H. P., 1970, The evolution of closed basin brines: Mineralogical Society of America Special Paper 3, p. 273–290.

Holmes, A., 1965, Principals of physical geology: New York, Ronald Press, 1288 p.

Hurst, H. E., and Phillips, P., 1931, The Nile Basin. General description of the basin: Cairo, Government Press, Physical Department Paper.

Jones, B. F., 1966, Geochemical evolution of closed basin waters in the western Great Basin, Northern Ohio: Northern Ohio Geological Society, Proceedings, Second Symposium on Salt, Cleveland, Ohio, v. 1, p. 181–200.

Jones, B. F., Eugester, H. P., and Rettig, S. L., 1977, Hydrochemistry of Lake Magadi basin, Kenya: Geochimica et Cosmochimica Acta, v. 41, p. 53–72.

Kendall, R. L., 1969, An ecological history of the lake Victoria basin: Ecological Monograms, v. 39, p. 121–176.

Kheiralla, M. K., 1966, A study of the Nubian Sandstone Formation of the

Nile Valley between 14°N and 17°42′N with reference to groundwater geology [M.Sc. thesis]: Khartoum, Sudan, Khartoum University, p. 235.

Lawson, A. C., 1927, The valley of the Nile: University of California chronicle no. 29.

Livingstone, D. A., 1980, Environmental changes in the Nile headwaters, *in* Williams, M.A.J., and Faure, H., eds., The Sahara and the Nile: Rotterdam, Balkema, p. 339–359.

Malmberg, G. T.,and Abdel Shafie, M., 1975, Application of environmental isotopes to selected hydrologic studies in Sudan: Vienna, International Atomic Agency, (Unpublished), 23 p.

Mcdougall, I., Morton, W. H., and Williams, M.A.J., 1975, Age and rates of denudation of Trap series basalts at Blue Nile gorge, Ethiopia: Nature, v. 254, p. 207–209.

National Council for Research, 1982, Water resources in Sudan: Khartoum, The National Council for Research, 233 p.

REGWA, 1979, Ed Dueim area, Hydrogeological study. Final report: Khartoum, Report submitted to Rural Water Corporation. (Unpublished), 94 p.

Rodis, H. G., Hassan, A., and Wahadan, L., 1963, Availability of groundwater in Kordofan Province, Sudan: Geological Survey of Sudan Bulletin, no. 12, 16 p.

Ruxton, B. P., 1965, The major rock groups of the northern Red Sea Hills, Sudan: Geological Magazine, v. 3, p. 314–330.

Said, R., 1981, The geological evolution of the River Nile: New York, Springer-Verlag, 151 p.

Salama, R. B., 1985a, Buried troughs, grabens and rifts in Sudan: Journal of African Earth Sciences, v. 3, p. 381–390.

Salama, R. B., 1985b, The evolution of the River Nile, in relation to buried saline rift lakes and water resources of Sudan [Ph.D. thesis]: Kensington, N.S.W., Australia University of New South Wales, 426 p.

Salama, R. B., 1987, The evolution of the River Nile. The buried saline rift lakes in Sudan—I. Bahr El Arab rift, the Sudd buried saline lake: Journal of African Earth Sciences, v. 6, p. 899–913.

Salama, R. B., 1993, The rift basins of Sudan, *in* Hsü, K., ed., Sedimentary basins of the World: Elsevier (in press).

Salama, R. B., and Salama, M. N., 1974, Geology and hydrogeology of the southern Sudan: Khartoum, Rural Water Corporation Open-File Report. (Unpublished), 22 p.

Schull, T. J., 1988, Rift basins of interior Sudan: Petroleum exploration and discovery: American Association of Petroleum Geologists Bulletin, v. 72, p. 1128–1142.

Shukri, N. M., 1949, The mineralogy of some Nile sediments: Quarterly Journal of the Geological Society of London, v. 105, p. 511–531.

Sly, P. G., 1978, Sedimentary processes in lakes, *in* Lerman, A., ed., Lakes: New York, Springer-Verlag, p. 65–84.

Talling, J. F., 1957, The longitudinal succession of water characteristics in the White Nile: Hydrobiologia, v. 11, p. 73–89.

Talling, J. F., and Talling, J. B., 1965, The chemical composition of African lake waters: International Revue der Gesamten Hydrobiologie, v. 50, p. 421–463.

Vail, J. R., 1978, Outline of the geology and mineral deposits of the Democratic Republic of Sudan and adjacent areas: Overseas Geology and Mineral Resources, no. 49, 66 p.

Wendorf, F., and Schild, R., 1976, Prehistory of the Nile Valley: New York, Academic Press, 404 p.

Whiteman, A. J., 1971, The geology of the Sudan Republic: London, Clarendon Press, 290 p.

Wigley, T.M.L., 1977, WATSPEC. A computer program for determining the equilibrium speciation of aqueous solutions: British Geomorphological Research Group, Technical Bulletin 20, 48 p.

Willcocks, W., 1904, The Nile in 1904: London, E. and F.N. Spon., 225 p.

Williams, M.A.J., 1966, Age of alluvial clays in the western Gezira: Nature, v. 211, p. 270–271.

Williams, A. J., and Adamson, D. A., 1973, The physiography of the central Sudan: Geographical Journal, v. 139, p. 498–508.

Williams, M.A.J., and Adamson, D. A., 1980, Late quaternary depositional history of the Blue and White Rivers in central Sudan, in Williams, M.A.J., and Faure, H., eds., The Sahara and the Nile: Rotterdam, Balkema, p. 281–304.

Williams, M.A.J., and Williams, F., 1980, Evolution of Nile basin, in Williams, M.A.J., and Faure, H., eds., The Sahara and the Nile: Rotterdam, Balkema, p. 207–224.

Yuretich, R. F., and Cerling, T. E., 1983, Hydrogeochemistry of Lake Turkana, Kenya: Mass balance and mineral reactions in an alkaline lake: Geochimica et Cosmochimica Acta, v. 47, p. 1099–1109.

Manuscript Accepted by the Society July 2, 1993

Printed in U.S.A.

Geological Society of America
Special Paper 289
1994

Carbonate and evaporite sedimentation at Clinton Lake, British Columbia, Canada

Robin W. Renaut
Department of Geological Sciences, University of Saskatchewan, Saskatoon, Saskatchewan, S7N 0W0 Canada

ABSTRACT

Clinton Lake is a small ephemeral lake lying in a glacial paleomeltwater channel near Clinton, Interior British Columbia. The lake is a Mg-Na-SO_4 brine (>300 g l^{-1} TDS), similar to those at Basque, 50 km to the south. Like the Basque Lakes, Clinton Lake has permanent deposits of epsomite.

There are six main depositional subenvironments: (1) hillslopes, (2) springs, (3) carbonate playa, (4) hummocky carbonate mudflats, (5) saline mudflats, and (6) ephemeral lake or saline pan. There is no channelled inflow, and Clinton Lake is fed mainly by groundwater, direct precipitation, and diffuse runoff. About 120 m north of the lake, at slightly higher elevation, lies a small carbonate playa, called Clinton Pond. This is fed by a series of small springs and seepages. Other springs issue from the basin margins, but most seep into the ground or evaporate before reaching the lake. Shallow brine covers Clinton Lake during spring and early summer, but most of the surface is salt encrusted by late summer and autumn.

Preliminary analyses of the basin waters and sediment mineralogy suggest that Clinton Lake may be fed by at least two distinct groundwater sources. The first type has a very high Na/Cl ratio (>15); the second type is less evolved, less saline, and has a lower Na/Cl ratio (<12). Both types of groundwaters have very high Mg/Ca ratios, resulting from contact with Mg-rich bedrock and early precipitation of calcite cements in soils.

Aragonite and Mg-calcite are precipitated at the spring orifices, further depleting the waters in calcium. Some spring waters flow into Clinton Pond, where aragonite and hydromagnesite precipitate. Pond waters, augmented by other springs, may seep below hummocky carbonate mudflats toward Clinton Lake. At Clinton Lake, these waters mix with other shallow, dilute seepage and possibly, artesian groundwaters discharging through the lake floor.

Cores and surficial muds from the saline mudflat contain hydromagnesite, aragonite, dolomite, and magnesite, and gypsum. Most calcium and much of the carbonate are precipitated in the outer mudflat, leaving a Mg-Na-SO_4 brine. Clinton Lake is a "spotted lake," consisting of more than 100 brine pools separated by rims of Mg-carbonate and gypsum muds. In summer, clear epsomite crystallizes in the pools as rafts and bottom-nucleated crystals. Mirabilite forms from residual brine in the autumn. Permanent epsomite deposits, showing evidence for repeated dissolution and recrystallization, underlie the pools.

Clinton Lake is the successor to an earlier Mg-carbonate marl lake. Historical records suggest that evaporite precipitation could be very recent.

Renaut, R. W., 1994, Carbonate and evaporite sedimentation at Clinton Lake, British Columbia, Canada, *in* Rosen, M. R., ed., Paleoclimate and Basin Evolution of Playa Systems: Boulder, Colorado, Geological Society of America Special Paper 289.

INTRODUCTION

British Columbia has more than a thousand saline lakes, most of which lie on the semiarid Interior Plateau between the Coast Mountains and Columbia–Rocky Mountain ranges. Although very small compared to those in the western United States, the lakes are diverse, both in type and in chemical composition. They range from small playas with mainly siliciclastic and carbonate sediments, to perennial saline lakes and saline pans. The saline lakes annually precipitate a wide variety of carbonates and evaporite minerals, including natron, mirabilite, and epsomite (Reinecke, 1920; Cole, 1924; Goudge, 1926a, b; Cummings, 1940; Renaut and Long, 1987, 1989; Renaut et al., 1994).

The largest group lies in the Cariboo-Chilcotin region of the south-central Interior Plateau, with the greatest concentration on the southern Cariboo Plateau, near Clinton and 70 Mile House (Fig. 1). In a region of <500 km^2, more than 500 saline and hypersaline lakes with Na-CO_3-(SO_4)-Cl, Na-CO_3-Cl, Na-Mg-SO_4, and Mg-Na-SO_4 brines are found. In the same region, there are also many ephemeral lakes with predominantly magnesium carbonate sediments, including abundant magnesite, hydromagnesite, and recent dolomite (Cummings, 1940; Grant, 1987; Renaut and Stead, 1991; Renaut, 1993).

The purpose of this paper is to present a preliminary description of the sediments and mineralogy of Clinton Lake, a small saline pan with a Mg-Na-SO_4 brine that is precipitating epsomite ($Mg_2SO_4.7H_2O$). In composition, the brine and mineralogy resemble the well-known Basque Lakes, near Ashcroft 50 km to the south (Fig. 1; Cole, 1924; Goudge, 1926a; Eugster and Hardie, 1978; Nesbitt, 1974, 1990; Renaut and Stead, 1994a). Like several of the Basque Lakes, Clinton Lake is a "spotted lake," consisting of more than 150 individual brine pools that are probably fed by upwelling groundwater and are separated by mud ridges. The Clinton Lake catchment, however, is much larger than those of the Basque Lakes and its basin has a much wider range of depositional subenvironments.

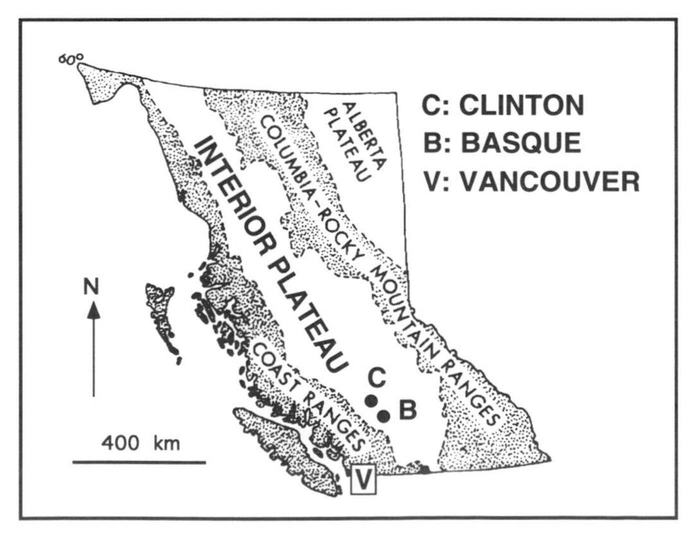

Figure 1. British Columbia, showing location of Clinton and the Basque Lakes.

The occurrence of epsomite at Clinton Lake was first described by Reinecke (1920), later supplemented by the observations of Cole (1924), Goudge (1926a), and Cummings (1940). Water analyses were reported by Topping and Scudder (1977). Small quantities of epsomite of high purity were extracted by pick and shovel between 1918 and 1920 (Goudge, 1926a), and in the 1940s. Although economic interest has been maintained, including small-scale extraction of lake muds for dermatological purposes in the 1970s, there have been no previous systematic studies of the sediments and their mineralogy.

METHODS

Fieldwork was undertaken at Clinton Lake in May to June 1985, with further brief visits in June 1986, October 1987, August 1991, and June to July 1992. Sediments were collected as grab samples at the surface, from shallow (<1 m) pits, and from short cores (3.8 cm diameter, <1 m long). Mineralogy was determined by X-ray diffraction (XRD) and petrographic methods. Several sediment samples were examined by scanning electron microscopy (SEM), using a JEOL JXA 8600 electron microprobe, equipped with wavelength and energy-dispersive spectrometers. Methods used for water analyses are given in Renaut (1990). The theoretical state of saturation of the waters with respect to various mineral phases was calculated using the PC version (Rollins, 1989) of the WATEQ4F program (Ball et al., 1987).

Many of the salts (e.g., mirabilite) are very sensitive to temperature changes and are unstable in the atmosphere. Epsomite, for example, begins to dehydrate rapidly to hexahydrite ($MgSO_4.6H_2O$) when removed from the brine. Although samples were stored in sealed plastic bags and original brine, some salts underwent change. Where possible, original mineralogy was reconstructed using (1) physical properties in the field, (2) X-ray diffraction of moist and dry samples, and (3) published data indicating which phases are likely to be stable at the temperatures measured at time of collection.

GEOLOGICAL AND ENVIRONMENTAL SETTING

Clinton Lake lies 2 km south of Clinton village, approximately 230 km northeast of Vancouver (Fig. 2). Clinton is located at the eastern edge of the Coast Mountains near their junction with the gently undulating Interior Plateau of British Columbia. The lake occupies a small closed basin in steep hilly terrain at the northern end of Alkali Valley, a narrow (100 to 500 m wide), structurally controlled, glacial paleomeltwater channel trending north-northwest to south-southeast. One kilometer north of Clinton Lake, Alkali Valley joins a larger, east-northeast to west-southwest trending paleomeltwater channel called Cutoff Valley, which is drained by Clinton Creek.

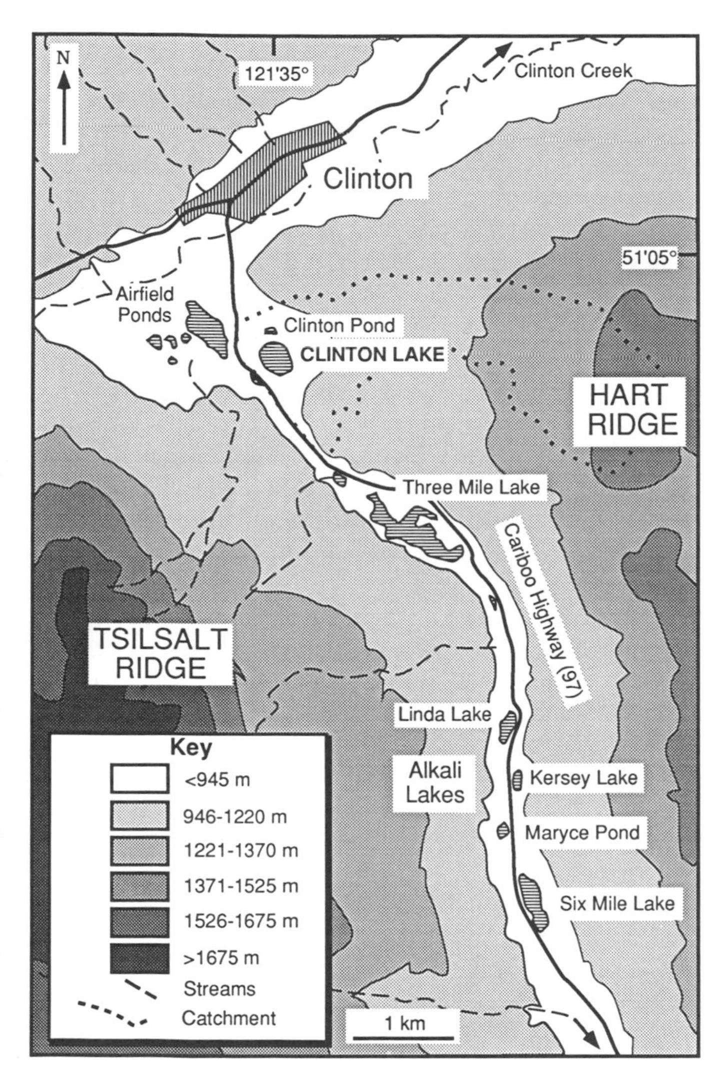

Figure 2. Alkali Valley, showing regional setting of Clinton Lake and its catchment.

The Coast Mountains at this latitude are represented by the Marble Range—part of the Cache Creek accretionary terrane of Carboniferous-Jurassic age (Monger, 1989; Monger et al., 1991). In the Tsilsalt Ridge (1,860 m) west of Clinton Lake and in Hart Ridge (1,400 m) to the east, the Cache Creek rocks are mainly argillites, massive limestones, cherts, basic lavas, graywackes, and shales (Campbell and Tipper, 1971), many of which are strongly sheared and fractured. The northern part of Alkali Valley, including much of the Clinton Lake drainage basin, is floored by very poorly exposed siliciclastic sediments of unknown thickness, assigned by Campbell and Tipper (1971) to the Deadman River Formation of Miocene age. These are largely undeformed, fluviatile conglomerates and sands, with thin interbedded lacustrine diatomaceous silts.

Although the entire region experienced multiple glaciation (Fulton, 1984; Clague, 1981; Clague et al., 1989), local till cover is very thin (0.5 to 5 m). Deglaciation took place ~10,000 years ago, producing enormous volumes of meltwater that flowed southward via the Thompson and Fraser drainage systems. Patches of imbricated glaciofluvial gravels are found on slopes up to at least 25 m above the floors of both Alkali and Cutoff valleys. The valley floors are mantled discontinuously by post-glacial colluvial gravels, sands, and silts up to a few meters thick, locally with one or more thin (<10 cm) Holocene tephra horizons. The glaciofluvial and post-glacial deposits contain a high percentage of volcanic rock fragments, derived mainly from the Interior Plateau to the northeast.

Clinton Lake is one of eight small closed saline lakes in Alkali Valley. There is very little surface drainage along the marshy valley floor, and all the lakes derive most of their recharge from groundwater, supplemented by minor runoff and direct precipitation. A north-south drainage divide is present south of Three Mile Lake (Fig. 2). Except for Clinton Lake and the northern subbasin of Three Mile Lake, which also dries out annually, all the lakes are perennial. Only Clinton Lake and Three Mile Lake are known to have evaporites; most of the others precipitate carbonate sediments that have not yet been studied.

The climate is semiarid, and characterized by warm summers and very cold winters. The mean annual precipitation is about 400 mm, a total similar to the mean annual moisture deficit (Valentine and Schori, 1980). Mean July temperatures range from 13 to 17°C, compared with –9 to –11°C in January (Atmospheric Environment Service, 1982). Snow and ice normally blanket the region from November until mid-March. The region lies within the "Interior Douglas Fir" biogeoclimatic zone (Annas and Coupe, 1979). The hillslopes have mixed Douglas fir, Lodgepole pine, aspen, white spruce, and Rocky Mountain juniper; the valley floors, including that surrounding Clinton Lake, are predominantly open grasslands with sagebrush. Further details are given in Valentine and Schori (1980) and Renaut and Long (1989).

Clinton Lake lies beside the Cariboo Highway (B.C. Highway 97), which is routed through Alkali Valley. However, the lake lies on private land that is used for ranching and permission must be sought to gain access.

DRAINAGE BASIN AND HYDROLOGY

The catchment (Fig. 2), with an area of ~5 km^2, is bounded to the east, north, and south by the western slopes of Hart Ridge. The western margin is represented by the embankment of the Cariboo Highway, which follows the western edge of the valley. Although the highway has modified the drainage, the lake occupies a natural depression that predates road construction.

Clinton Lake (Fig. 3) is oval, with elongation parallel to the valley axis. The lake lies at ~910 m above sea-level and has a maximum diameter of ~400 m. There is no channelled inflow and the lake is fed by groundwater, diffuse runoff, and direct precipitation on the lake's surface (including snowmelt). Most

Figure 3. View east across Clinton Lake pan toward Hart Ridge. Photograph taken in March during snowmelt. Saline pan surface is covered by crust of epsomite and mirabilite.

precipitation falling on the catchment is absorbed by the forest vegetation or seeps rapidly into the sandy soils. Significant overland flow has only been observed while the ground is still frozen after snowmelt in March to April, but little reaches the lake as runoff. In most years, groundwater recharge and evaporation are the main factors in the hydrological budget. Groundwater flows into Clinton Lake from shallow ephemeral seepages around its perimeter and may also flow directly into the lake via a shallow aquifer below the lake floor.

In common with most ephemeral lakes in the region, water levels in Clinton Lake vary seasonally, and from year to year. Following snowmelt and early summer rains, the maximum depth may reach ~1 m. During the spring, small shallow rivulets (10 to 100 cm wide; <3 cm deep) flow intermittently across the lake marginal mudflats from seepage zones along the base of the hillslopes. During this phase, at least the upper few centimeters of the previous year's new salts are dissolved. By late June, little surficial flow remains, but the soft lake marginal muds are saturated close to the surface by moderately fresh to brackish waters (~2 to 10 g l^{-1} TDS; total dissolved solids). With reduced inflow and increased evaporation during midsummer, lake level falls and salts begin to crystallize from the dense, brownish brine. By August the surface of the pan is commonly a white to yellowish white, dry salt crust. Rains falling during late summer and autumn may partly dissolve the crusts. Any surface water remaining by October eventually freezes and the lake becomes snow and ice covered until the next spring thaw. Similar annual cycles are recorded in many ephemeral saline lakes in western Canada (Last, 1984, 1989; Renaut and Long, 1987, 1989; Renaut and Stead, 1994b). Much of the more soluble evaporites that precipitate in summer are recycled annually.

Reinecke (1920, p. 53) commented that, "The quantity of water in the lake seems to vary daily even during continued periods of rainless weather." While this could be related to diurnal changes in pressure, this phenomenon has also been observed at times when air temperatures drop close to zero by night, especially during late spring and autumn. With falling temperature, ephemeral crystal rafts up to 2 m across crystallize at the air-water interface. The rafts, which are composed of narrow interlocking bladed crystals up to 15 cm in length, are only a few millimeters thick. With daytime warming, the salts dissolve releasing water and lake-level rises slightly or, if partially dry, the surface is reflooded by thin film of brine. XRD analyses have shown that both mirabilite ($Na_2SO_4.10H_2O$) and epsomite rafts can form in this way. Mirabilite rafts seem to be more common in spring. Similar, temperature-controlled diurnal variations in water level occur in local sodium carbonate lakes, where natron ($Na_2CO_3.10H_2O$) crystallizes (Renaut and Long, 1987), and in many sodium sulfate lakes in Saskatchewan (Last, 1984).

The general setting of Clinton Lake is shown in Figure 4. Approximately 100 m north of Clinton Lake there is a small carbonate playa, named here Clinton Pond, which is approximately 100 m long by 50 m wide, and when covered by water has an annual maximum depth of ~50 cm. Clinton Pond occupies a small subbasin approximately 1 m above the normal maximum level of Clinton Lake. It also is usually dry by late summer, but in some years (e.g., 1990), water remains throughout the year.

Clinton Pond is fed mainly by subaerial springs whose waters flow into the playa, but may also be augmented by direct groundwater discharge through the lake floor. The largest spring, Clinton Lake Spring, discharges at 7 to 8°C from a pool in ~3 m in diameter and 1 m deep, located ~30 m north of the playa lake. It flows through a narrow (1 m wide, 20 cm deep) marshy channel into the northeastern corner of Clinton Pond

Figure 4. View southward across Clinton Pond toward Clinton Lake. The playa is beginning to dry out. A spring issues from the reed zone on the left.

Figure 5. Zone of springs and seepage northeast of Clinton Pond (visible center left). Clinton Lake Spring is to the right of the figures; other springs seep from the marshy ground on the left. Note extensive carbonate efflorescence.

(Fig. 5). About eight small springs issue from marshy ground 10 to 40 m east and southeast of Clinton Lake Spring. Some flow into Clinton Pond; most seep into damp marshy ground a few meters from their source. Another spring seeps directly into a dense reedy zone on the floor of north-central Clinton Pond (Fig. 4). All these springs have their maximum discharge from April until late June. Most are ephemeral and dry up by late summer, but Clinton Lake Spring appears to flow throughout the year, even below an ice cover 10 to 15 cm thick.

HYDROCHEMISTRY

Representative compositions of the basin waters, together with analyses of several of the local Alkali Valley lakes (Fig. 2), which lie on the same bedrock, are given in Table 1. The most dilute waters were collected from shallow surface rills draining basin slopes after six hours of heavy rainfall, and from Cutoff Creek at Clinton. Although not within the catchment, the latter sample represents runoff draining the Marble Range.

Clinton Lake Spring and the small seeps to the east (Seeps 1 to 3 in Table 1) are moderately fresh (2.5 to 3 g l^{-1} TDS). Their chloride concentrations range from two to six times higher than local runoff, but their chemical composition suggests that they are not simply the result of evaporative concentration of the latter waters. The molar Na/Cl ratio has increased by up to a factor of 11, and the Mg/Ca ratio has increased from <1 to >60.

Clinton Pond has the highest values of pH recorded in the basin (7.5 to 9.2), and has a measured salinity that ranges from <2 g l^{-1} TDS in spring to >15 g l^{-1} TDS in late summer. The Na/Cl ratio is similar to the adjacent springs, but the molar Mg/Ca ratio, when sampled, exceeded them by a factor of between 2 and >5.

The shallow groundwaters and surface seepages around the shoreline of Clinton Lake range from fresh to moderately saline (1 to 12 g l^{-1} TDS), and vary in their compositions. Water samples collected from ephemeral seepages along the southeastern shores (CB-13 and 14) have Na/Cl ratios similar to those of the springs north of Clinton Pond. In contrast, waters sampled from a shallow pit dug in peripheral mudflats at the southwestern shoreline (BCS-12) and from surficial seepage at the northeastern shoreline (BCS-10) have higher Na/Cl ratios. In common with the springs, very high Mg/Ca ratios are found in both groups.

Clinton Lake brine is Mg-Na-SO4 in composition, with a pH of 8.0 to 8.3 and a salinity that ranges annually from <100 g l^{-1} TDS in spring to >350 g l^{-1} TDS in late summer. The pH of surface waters does not appear to vary significantly throughout the year. Measured brine temperatures range from 10 to 12°C in open brine during May to 35°C below salt crusts in August. Maximum recorded salinities are in mid-to-late summer, but there are no data for winter. In other Cariboo saline lakes, such as Goodenough Lake, maximum salinity may occur in midwinter below an ice cover (Topping and Scudder, 1977). The limited data available suggest that the Na/Cl ratio of the brine varies at different times of the year, reaching a maximum in the late spring following snowmelt and associated groundwater recharge, and a minimum in late summer and autumn following precipitation of sodium sulfate and chloride salts.

Biological studies have not been undertaken at Clinton Lake. Hammer and Forró (1992) have recorded the zooplankton from neighboring Three Mile Lake. *Artemia* sp., various invertebrate larvae, and cyanobacterial mats have been observed in Clinton Lake, mostly during spring and summer.

SEDIMENTS AND MINERALOGY

Clinton Lake is a saline pan bounded by narrow saline mudflats and dry, grass-covered, mixed carbonate-siliciclastic mudflats. The modern sediments of the pan consist mainly of highly soluble magnesium and sodium evaporitic salts, mixed and interlayered with sparingly soluble Mg–carbonates, gypsum, organic matter, and detrital siliciclastic material. The small playa is dominated by carbonates, while the mudflats surrounding Clinton Lake are variably composed of carbonates, gypsum, and slope-derived siliciclastics.

The lake basin can be divided into six main subenvironments (Fig. 6), each of which has a distinctive group of microfacies: (1) hillslope, (2) springs, (3) carbonate playa, (4) hummocky carbonate mudflats, (5) peripheral saline mudflats, and (6) saline pan. Figures 7 and 13 show representative sediment profiles for each of the main subenvironments.

Hillslopes

Most slopes are moderately well stabilized by vegetation, but locally show small curvilinear arcuate failures. Small terracettes are common on slopes of 5 to 15° in the grass-covered

TABLE 1. CHEMICAL ANALYSES OF CLINTON BASIN WATERS

Sample	Description	Date	pH	Na (mg/l)	K (mg/l)	Ca (mg/l)	Mg (mg/l)	Cl (mg/l)	Alkalinity (meq/l)	SO_4 (mg/l)	F (mg/l)	SiO_2 (mg/l)	Na/Cl (mol)	Na/K (mol)	Mg/Ca (mol)
Surface runoff															
CT-1	Cutoff Creek	Aug-91	8.20	2.9	0.75	71	14	4.2	4.10	28	0.2	10	1.06	6.58	0.32
BCS-5	Hart Ridge runoff-a	Jun-92	7.75	9.6	1.66	45	13	2.9	3.03	31	0.22	8	5.11	9.84	0.48
BC-11	Hart Ridge runoff-b	Aug-87	7.95	9.1	2.15	51	26	3.5	4.26	42	n.d.	11	4.01	7.20	0.84
BCS-6	Tsilsalt Ridge runoff	Jun-92	7.30	4.2	0.95	56	16	3.9	3.44	32	0.45	5	1.66	7.52	0.47
Springs															
CB-6	Seep 1	Jun-85	7.30	75	14	12	450	11	26.22	610	n.d.	34	10.52	9.11	61.80
CB-9a	Seep 2	Jun-85	7.20	65	13	22	420	13	23.76	640	n.d.	34	7.71	8.50	31.46
BCS-8	Seep 3	Jun-92	7.25	70	11	28	395	9	24.58	595	0.6	36	11.99	10.82	23.25
CT-2	Clinton Lake Spring	Aug-91	7.30	85	20	16	440	16	26.22	600	n.d.	40	8.19	7.23	45.32
BCS-7	Clinton Lake Spring	Jun-92	7.20	72	11	30	410	12	27.53	510	0.8	36	9.25	11.13	22.52
Clinton Pond															
BC-4	Clinton Pond	Jun-85	9.20	136	24	4	540	23	26.53	1,235	n.d.	11	9.12	9.64	222.48
CT-3	Clinton Pond	Jun-86	9.00	155	27	5	461	26	24.81	1,010	n.d.	3	9.19	9.76	151.95
BCS-8	Clinton Pond	Jun-92	9.20	195	35	3	575	31	32.68	1,095	1.9	5	9.70	9.48	315.87
Lake marginal waters															
BCS-10	Groundwater, N.E. shore	Jun-92	8.10	390	56	8	460	36	28.05	1,340	1.8	13	16.71	11.84	94.76
BCS-12	Groundwater, S.W. shore	Jun-92	8.50	1,450	145	9	1,365	85	26.55	7.290	2.3	11	26.31	17.01	249.95
CB-13	Seep, S.E. shore	Jun-86	7.80	335	40	5	960	42	33.50	2,800	n.d.	11	12.30	14.24	316.42
CB-14	Seep. S. shore	Jun-86	7.50	40	8	15	110	6	10.00	80	n.d.	27	10.28	8.50	12.09
Clinton Lake															
BC-24	Clinton Lake Brine	May-85	8.10	19,600	1,360	30	42,500	1,400	97.32	198,000	n.d.	21	21.50	24.51	2,335
CB-16	Clinton Lake Brine	Jun-86	8.20	21,800	1,750	45	61,500	3,550	73.75	275,000	n.d.	10	9.47	21.19	2,252
BCS-17	Clinton Lake Brine	Jul-92	8.25	25,600	3,050	55	46,500	3,750	98.33	289,000	11.6	22	10.53	14.27	1,855
Alkali Valley Lakes															
BC-17	Airfield Pond A	Aug-88	9.10	225	42	0	450	21	24.65	1,080	3.8	21	16.52	9.11	
BC-14	Three Mile Lake	Aug-88	8.60	2,945	250	185	6,500	130	38.13	30,860	4.2	16	34.94	20.03	57.90
BCS-18	Three Mile Lake	Jun-92	8.70	7,010	560	210	14,350	315	50.21	69,450	10	21	34.32	21.29	112.62
BC-19	Pond B	Aug-88	8.80	125	11	7	165	14	7.38	560	2.1	15	13.77	19.33	38.85
BCS-20	Pond A	Jun-92	8.30	87	8.6	26	120	11	7.46	340	1.9	13	12.20	17.20	7.61
BC-21	2.5 Mile Lake	Jul-85	8.70	63	10	30	250	9.5	11.62	650	2.3	4.2	10.23	10.71	13.73
BC-31	Kersey Lake	Jul-85	8.40	210	16	355	450	80	2.54	2,850	1.7	4.5	4.05	22.32	2.09

n.d. = not determined.

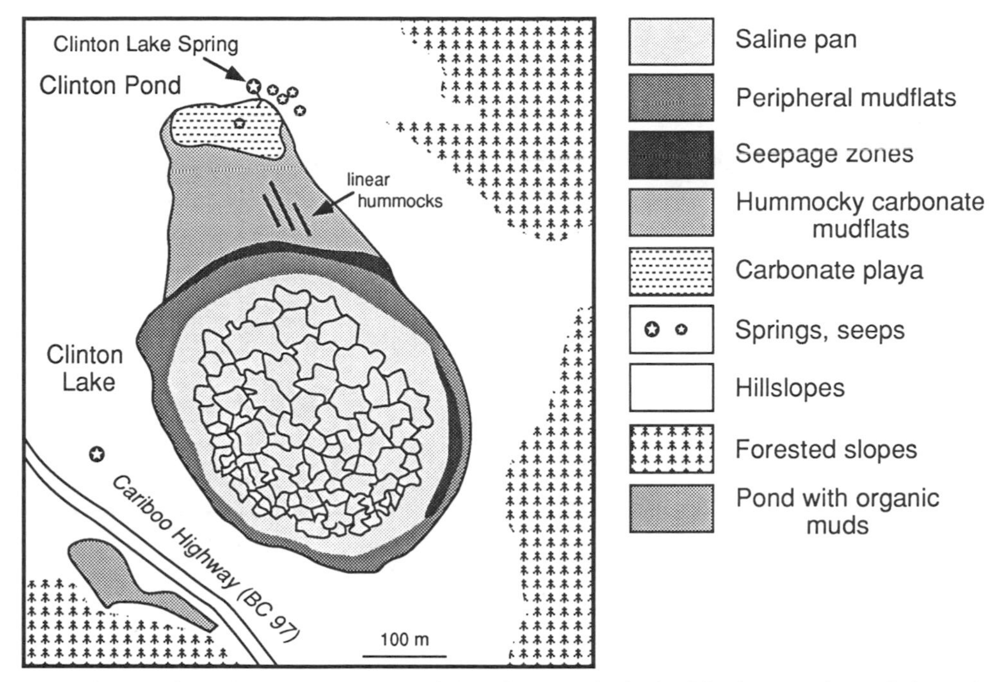

Figure 6. Depositional subenvironments of the Clinton Lake basin. The large-scale mud-rimmed structures shown in the saline pan are drawn from aerial photographs. The brine pools lie *within* these larger structures.

zone surrounding Clinton Lake. Most soil creep and mass wasting occur in spring, following thawing. The clastic sediments derive mainly by sheetwash from the adjacent slopes, but very little sediment released by frost heave reaches the lake or the playa.

The surficial sediments on hillslopes are mainly poorly sorted, grayish brown, silty and gravelly sands, and sandy silts. They are variably composed of quartz, K-feldspar, plagioclase, mica, and lithic grains (basalt, argillite, shale, chert, limestone, and fine-grained metamorphic rocks). Most hillslope colluvium is thin (<30 cm).

Powdery, intergranular micritic calcite is common in the upper 15 to 20 cm, generally increasing in abundance with proximity to the two lakes. The undersides of gravel clasts in the upper 30 cm also commonly have thin (1 to 2 mm), white micritic carbonate rinds, probably representing gravitational cements. XRD data show that both low-Mg and high Mg-calcite (up to 17 mol% $MgCO_3$) are present, the latter mostly within 10 m of the shoreline of Clinton Lake. On the lower western slopes of Hart Ridge, fine white powdery gypsum accompanies Mg-calcite as an interstitial precipitate in the surficial sediments. The authigenic calcite and gypsum are mostly products of evaporative concentration and evapotranspiration of shallow soil waters.

Springs

Alkaline earth carbonates are precipitated around most of the spring orifices, but well-developed travertine deposits are absent. The land surface surrounding the damp, marshy ground at seepage sites is usually covered by a thin veneer of white or gray, fine powdery carbonates (Fig. 5). XRD analyses show that most are either aragonite or hydromagnesite, with minor Mg-calcite and calcite. The carbonates occur as (1) thin (<1 cm), puffy, discontinuous crusts around and downstream from the orifice; (2) thin (1 to 4 mm), weakly laminated, surficial encrustations on exposed gravel clasts embedded in surficial sediments; or (3) loose, powdery efflorescence. The powdery efflorescence occurs both at the surface (up to 0.5 cm thick) and within intergranular porosity of the upper few centimeters of host sediment. Filamentous microbial mats are associated with some crusts, notably on gravel clasts at the orifices, but few well-laminated microbolites (sensu Riding, 1991) are found. Poorly crystalline goethite, and X-ray amorphous iron and manganese oxides also weakly cement siliciclastic sediments around many springs. Orange colloidal iron oxides and hydroxides also choke some outflow channels.

Clinton Lake Spring, being perennial, has somewhat different deposits. Carbonates are precipitated mainly as intergranular cements in the gravelly siliciclastic sediments that surround the spring orifice. No flowstones were recognized. The cemented zone extends laterally up to about 1.5 m from the orifice. Cementation is most intense at the surface, locally forming an indurated crust a few centimeters thick, and extends downward up to ~50 cm. There are two main generations of cement—an early phase of brownish micritic low-Mg calcite, and a later phase of clear needle-fiber aragonite that fills microfractures cutting the earlier calcite. A similar change from calcite to aragonite precipitation is found in several other

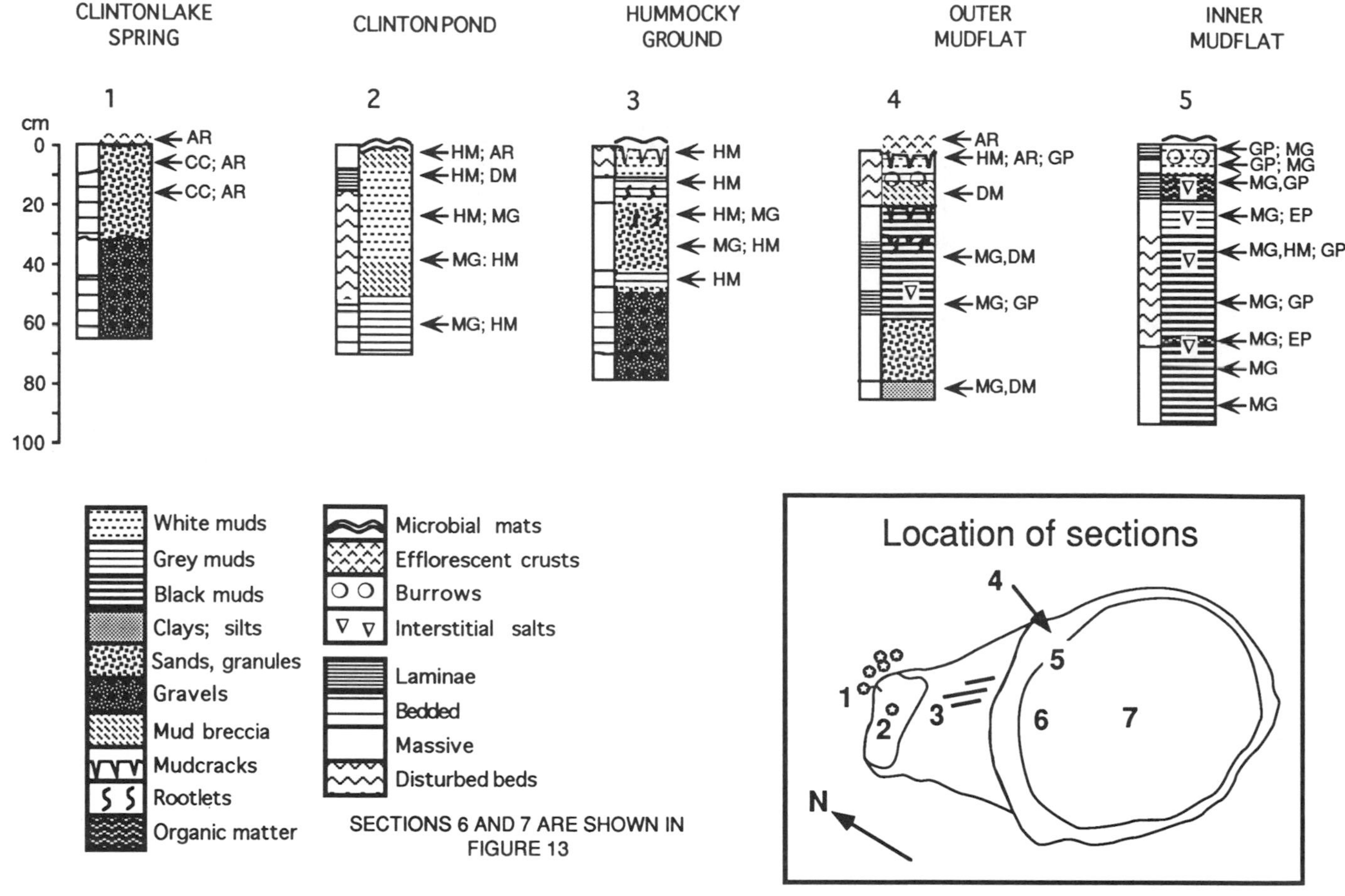

Figure 7. Selected profiles through sediments of Clinton Lake basin. For key to mineralogy see Figure 14.

local spring deposits. One possible explanation is a climatically induced increase in the Mg/Ca ratio of the spring waters through time (Renaut, 1990).

The spring carbonates are probably precipitated as a result of evaporative concentration in the damp capillary zone surrounding the orifices, degassing of CO_2, and mediation by microorganisms (cf. Julia, 1983; Chafetz and Folk, 1984; Heimann and Sass, 1989). Two factors may partly explain the absence of well-developed travertine or tufa. First, most of the springs are ephemeral or have substantially reduced winter discharge: surficial carbonates precipitated in spring and early summer are readily eroded by deflation when dry, or loosened by segregation ice in winter. Second, the high molar Mg/Ca ratios (22 to >60) favor rapid precipitation of friable needle-fiber aragonite or platy hydromagnesite, rather than less soluble calcite encrustations. Aragonite only appears to survive where protected as an intergranular cement.

Carbonate playa

Clinton Pond (Fig. 4) is a small carbonate playa, floored mainly by gray and brownish gray carbonate muds, which in the center are at least 70 cm thick. A narrow (3 m) fringe of siliciclastic sands and silts interfingers with the carbonates along the steeper northern and eastern margins. Most of the playa muds are massive or weakly bedded, but the upper centimeter from the central (i.e., annually water-covered) playa may show an indistinct lamination when dry. Horizons showing microbrecciation due to desiccation and/or cryogranulation are found in cores and pits. Cryogranulation is the process of breakage of the muds into small subcuboidal granules and peloids by ephemeral, platy, segregation ice (Renaut, 1993). The muds are composed mainly of hydromagnesite and aragonite, with subsidiary nonstoichiometric dolomite, magnesite, fine siliciclastic material (mainly quartz-silt and clays), and organic matter.

The surficial muds are mostly hydromagnesite and aragonite, and are found as (1) nonlaminated, relatively pure, monomineralic muds; (2) massive muds of mixed composition; and rarely (3) laminated muds, in which fine (1 to 2 mm) white aragonitic laminae (Fig. 8) alternate irregularly with gray hydromagnesite laminae (2 to 4 mm; Fig. 9). Hydromagnesite occurs throughout the subbasin, and is dominant in the marginal

Figure 8. Scanning electron microscopy (SEM) photomicrograph of acicular aragonite from Clinton Pond.

Figure 9. Scanning electron microscopy (SEM) photomicrograph of hydromagnesite mud from Clinton Pond.

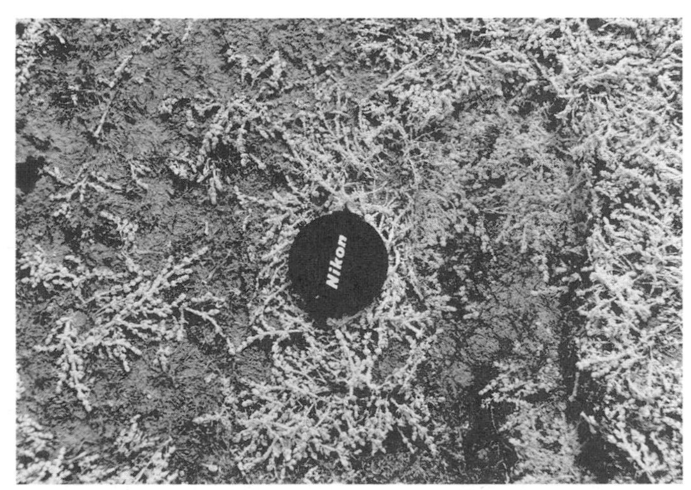

Figure 10. Charophytes encrusted by aragonite. Desiccated floor of Clinton Pond.

playa muds. Aragonite is most abundant on the eastern margins near the sites of spring water inflow, including that discharging directly through the lake floor. The laminated muds were found only in the east-central playa, near sites of freshwater seepage. Charophytes (Fig. 10), which are present near the sites of dilute inflow and seepage, are encrusted by aragonite rather than the more typical calcite. Microbial mats are present around the shoreline seeps and in wet years, thin (2 to 5 mm) benthic mats cover much of the lake bottom. The mats generally do not preserve in the playa muds.

Studies of hydromagnesite-magnesite playa lakes elsewhere in the region suggest that Mg-carbonates nucleate and precipitate in the lake annually, probably by evaporative concentration and microbial mediation (Renaut and Stead, 1991; Renaut, 1993). Although some aragonite could be detrital from local spring sites, its extracellular encrustation of charophytes and interlamination with hydromagnesite also support chemical precipitation from lake waters. Aragonite may precipitate when the playa lake is relatively fresh, eventually giving way to hydromagnesite as the Mg/Ca ratio, Mg-concentration, and salinity increase through evaporative concentration. The inflow of relatively fresh ephemeral seepage through the lake floor has produced a local inversion of the predicted carbonate mineral zonation—more aragonite is found in surface muds near the center of the playa than in the margins.

When desiccated, the playa floor becomes mudcracked, vesiculated, and thin crusts of efflorescent hydromagnesite and/or aragonite may cover the surface. Minor gypsum is commonly present. A short (70 cm) core of muds from the center of the playa revealed massive and weakly bedded hydromagnesite, overlying mixed hydromagnesite-magnesite muds at ~30 cm depth. A similar mineralogical change is found in other hydromagnesite-magnesite playas in the region, and may result from diagenetic alteration of hydromagnesite to magnesite (Cummings, 1940; Renaut, 1993). Minor, nonstoichiometric dolomite is found at depths of ~2 to 15 cm in aragonitic muds from the eastern playa margin and patchily elsewhere. Where present, the underlying muds are massive hydromagnesite, with little or no aragonite. Its variable distribution and association with aragonite suggest that the dolomite is replacive, but stable isotope analyses have not yet been made and the conditions of dolomite formation require investigation.

Hummocky carbonate mudflats

Carbonate mudflats occupy most of the area between Clinton Pond and Clinton Lake (Fig. 6). They are characterized by a low-lying hummocky surface with up to 40 cm of microrelief, covered by coarse grasses. Near Clinton Pond, the hummocks show well-defined polygonal patterns similar to those at Milk Lake, British Columbia (see illustrations in Renaut and Stead, 1991). The sediments in this zone are a mixture of carbonate muds and brown siliciclastic mud and sands. The surficial gray muds are predominantly hydromagnesite, which together with aragonite and minor gypsum, also occurs as surficial efflorescent crusts. The purest carbonate muds are found closest to Clinton Pond. The underlying sediments are typically massive pure carbonates, or a chaotic, mottled mixture of carbonates, siliciclastic materials, and organic matter. Most organic matter is present as roots (Fig. 7). Pure siliciclastic muds and sands are found at depths ranging from a few decimeters to a meter, according to location. Both sharp and transitional contacts have been observed in pits.

Hummocky surfaces are very common in carbonate playas of the Cariboo region. They are associated with sites of present or former capillary evaporation and upward movement of shallow groundwater. Hydromagnesite is precipitated interstitially in the sediments and within microbial mats, and, together with ephemeral segregation ice in winter, this may cause soil heave, raising the sediment surface (Renaut, 1993). Groundwaters in this zone lie within a few decimeters to a meter of the surface, being highest in the spring. Most muds near Clinton Pond were probably deposited subaqueously during former periods of lake expansion and have since been modified by groundwater and pedogenic processes.

Most hummocks are broadly circular, but parallel zones of elongate hummocks up to a meter wide and several meters long are found on the eastern part of the mudflat, north of Clinton Lake (Fig. 11). They extend southward toward a zone of groundwater seepage on the northeastern edge of Clinton Lake. Their elongate morphology may reflect a present or former zone of shallow southward groundwater movement between the two lakes (see below). They resemble the boxwork mounds associated with upwelling groundwaters illustrated by Warren (1982, Fig.8).

Figure 11. Linear hummocks of hydromagnesite, approximately 25 m north of Clinton Lake. View north toward Clinton Pond and springs (white area, top center).

Peripheral (saline) mudflats

A narrow (5 to 15 m), but continuous, mudflat surrounds the central saline pan of Clinton Lake. Two gradational subzones can be recognized based on facies and mineralogy: (1) an outer mudflat, which is only occasionally submerged by the lake and is composed mainly of carbonates and slope-derived siliciclastics; and (2) an inner saline mudflat that is submerged annually and contains significant amounts of gypsum.

Outer mudflat. The outer mudflat has a sharp contact with the hillslope. Although generally flat, low hummocky surfaces commonly mark local zones of shallow lakeward seepage. Patches of *Salicornia,* rhizomous salt-tolerant plants, and leathery microbial mats are present in these damp zones. In spring, shallow (1 to 3 cm) rills may flow briefly across the mudflat from peripheral seepage zones (Fig. 10 in Renaut and Long, 1989). The sediments (Fig. 7) are mainly dark gray and black, mottled carbonate muds, with stringers and lenses of clastics, including gravel clasts. The muds are either massive or show an indistinct coarse lamination. Other common features include rootlet horizons, subhorizontal burrows (3 to 5 mm diameter; 1 to 4 cm long), microbrecciation, mudcracks, vesicles, and microbial mat fragments, some of which are mineralized by hydromagnesite. In summer, powdery saline efflorescence and puffy crusts develop rapidly across the surface. Interstitial salts are then found locally in the upper few centimeters. Most sediments are reducing at 3 to 10 cm depth, yielding H_2S.

The carbonates in this zone are highly variable, mainly reflecting the presence or absence of local freshwater seepage, and subsequent diagenesis. Hydromagnesite and aragonite appear to be the most common carbonates in the surface muds, whereas magnesite and/or nonstoichiometric dolomite may accompany hydromagnesite in the shallow subsurface (>5 cm depth). The Ca-bearing carbonates are more common in near-surface sediments at zones of seepage. Fine gypsum is locally common, in many places exceeding the carbonate by volume. The surficial efflorescence is variably composed of hydromagnesite, aragonite, dolomite, or gypsum. Some hydromagnesite may have precipitated in a former, expanded, fresher Clinton Lake and may be altering to magnesite in the shallow subsurface. Some precipitation of hydromagnesite and aragonite also probably results from capillary evaporation of shallow groundwaters, as evident from hummocky and puffy ground.

Inner mudflat. The inner mudflat is generally flatter and for much of the year remains moist and extremely soft at the surface. Seepage waters locally emerge from near the inner margin of the inner mudflat with the saline pan. In a few places (e.g., northeastern shoreline), a narrow (1 to 2 m), dark rim of laterally continuous microbial mats becomes well developed just below water level. In wet years, the mats may remain submerged for several months; in drier years they appear not to grow. As the surface dries out, wrinkled efflorescent crusts up to 2 cm thick may form on the sediment surface. The loose powdery efflorescence characteristic of the outer mudflat zone is poorly developed or absent. Mudcracks are usually absent because the sediments remain continually moist, especially near sites of seepage.

The sediments in this zone are also variable in structure (Fig.7). Some are coarsely laminated; most are massive and mottled or have only indistinct lamination. Microbial laminites, in which nearly continuous black organic-rich layers (1 to 2 mm) alternate with carbonates and/or gypsum (2 to 4 mm thick), are found locally below sites where modern mats develop. Reducing conditions are commonly evident within 2 cm of the sediment surface and may account for the preservation of microbial lamination. This is the only zone where well-developed microbial lamination has been found in the sediments on the Clinton Lake basin. Burrows and rootlet horizons, some of which are rhizomes, are found in places in the muds at depths of up to 40 cm. Mudcracks and other desiccation features are poorly preserved or absent in the subsurface.

Like the outer mudflat, some of the sediments are carbonate muds with fine siliciclastics, but most contain large amounts of gypsum. The carbonates are mainly hydromagnesite or magnesite near the surface, with very fine grained magnesite and dolomite in the underlying black reduced muds. The gypsum occurs as (1) massive fine-grained muds that are either pure or mixed with magnesite and/or dolomite; (2) microbial laminites; and (3) scattered, coarse (5 to 10 mm) euhedral and subhedral prismatic crystals randomly oriented within the matrix of (1) and (2). The latter may be accompanied by epsomite crystals. The surficial efflorescence is mainly hydromagnesite and hexahydrite, but with local gypsum, thenardite (Na_2SO_4), aragonite, and halite.

The occurrence of scattered coarse, clear euhedral prismatic gypsum and epsomite crystals supports displacive growth from interstitial brines, but the origin of the fine-grained gypsum is unclear. There are several possibilities. It may precipitate directly from lake water during the phase of maximum expansion and dilution in the spring, perhaps where Ca-bearing inflow mixes with the strong sulfate brine. Alternatively, it may precipitate at or close to the sediment surface by evaporative concentration of shallow seepage waters, or it may be a relic deposit of a less saline paleolake. The occurrence of fine-grained gypsum-bearing microbial laminites favors precipitation directly from shallow lake waters.

Saline pan (ephemeral lake)

The surface of the saline pan of Clinton Lake is made up of numerous (>150) shallow pools of brine in which evaporites crystallize. These are separated from each other by rounded rims of mixed siliciclastic-carbonate-evaporite mud. The brine pools are broadly circular, but commonly have very irregular shapes, particularly toward the center of the lake, where they coalesce to form a complex interlocking jigsaw-like pattern. Individual pools range from 3 to ~10 m in diameter and when lake level is highest are up to ~1.5 m deep, increasing in both dimensions toward the center of the lake. Around the shoreline and in the southern third of the lake most pools are circular and are well separated by mud rims. In the northern, eastern, and central parts of the lake, groups of small mud-rimmed pools are themselves enclosed within larger mud-rimmed structures, some of which are more than 30 m across (Fig. 6). The interpool mud rims typically vary from 20 to 100 cm wide at the surface, increasing progressively in width with depth, suggesting that the pools are saucer to cone shaped in section. The widest mud rims, some of which exceed 3 m across, are found in the southeastern parts where the pools are correspondingly smaller.

Figure 12, based on observations made by Reinecke (1920) and excavations during this study, shows the general structure of the lake and its mineralogy. Because of the extremely soft conditions and consequent difficulties of excavation, no cores and pits have yet penetrated more than ~1.4 m below the surface crust. This interpretation is thus preliminary and the deeper structure awaits investigation.

Two individual pools, one located 20 m from the northern shore, the other near the center of the lake, were partially excavated in October 1987 when the surface sediments were dry and moderately firm. Both pools have a similar overall stratigraphy and mineralogy (Fig. 13). When dry, the bases of the pools have a 1- to 3-cm-thick surface layer of nearly pure, yellowish white to translucent clear epsomite. The crust consists mainly of subvertically aligned prismatic crystals, approximately 1 to 2 cm long by 0.3 to 0.8 cm wide, with less than ~5% visible intercrystalline porosity. The morphology is similar to the epsomite crusts illustrated by Pueyo and De la Peña (1991, Fig. 7A) from a saline lake in La Mancha, Spain. Many crystals have rounded corrosional terminations with overgrowths that probably result from partial dissolution by sum-

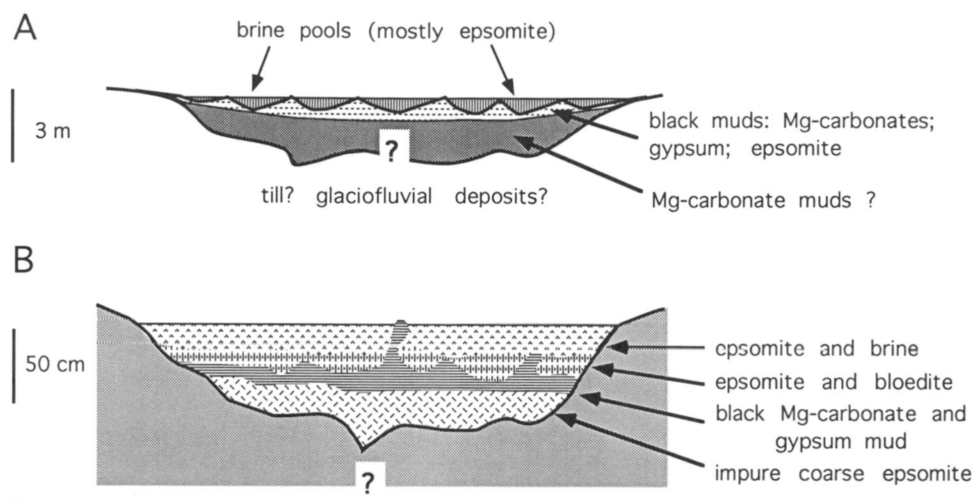

Figure 12. Hypothetical structure of Clinton Lake (modified from Reinecke, 1920). A, highly generalized structure across the saline pan; B, detailed stratigraphy of an individual brine pool.

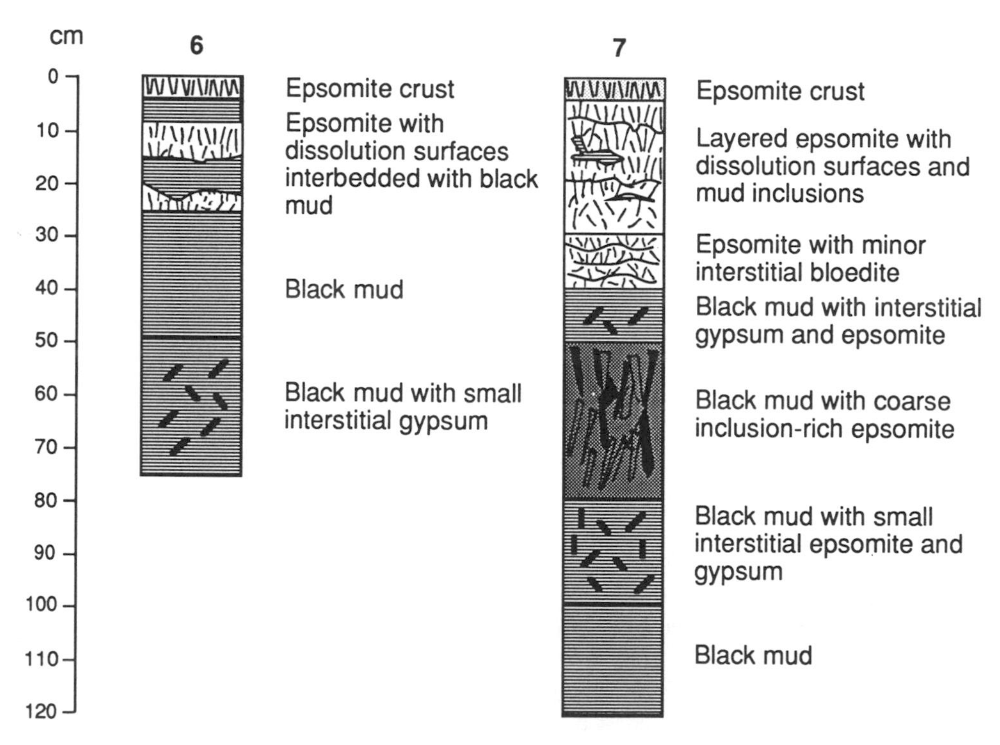

Figure 13. Stratigraphy of brine pool sediments in Clinton Lake pan. For location of sections, see Figure 7.

mer rains. Similar features are reported from many other saline pans (e.g., Lowenstein and Hardie, 1985; Warren, 1985; Last, 1989). Thin "dirt laminae" at the interface between some corroded crystals and their overgrowths are mostly microbial detritus or *Artemia* fecal pellets. These impart the yellow tinge to many crusts.

Within the crusts from the central pools there are also finer dense, subhorizontal, intraclastic crystal layers 5 to 10 mm thick and up to 2 cm long. Some intraclasts have an imbricate structure. The lower sides of some intraclasts have small, stubby, downward-pointing subhedral crystals, 2 to 5 mm long. These are interpreted to be foundered crystal rafts of epsomite that have sunk and been incorporated in the upward-growing crusts, analogous to the hopper rafts in halite pans

(Lowenstein and Hardie, 1985) and the radiating natron rafts in local sodium carbonate lakes (Goudge, 1926b; Renaut and Long, 1987). Epsomite rafts commonly form during midsummer and rapidly cover the entire brine pool surfaces both at Clinton Lake and in Basque Lakes Nos. 1 and 2. In July 1992, newly formed crystal rafts were observed to be piled up against mud rims on downwind margins of the larger brine pools Clinton Lake. These may explain the imbricate structures seen in some crusts. Large intraclasts in the crusts could also originate as overturned broken slabs from tepee ridges.

This surficial epsomite crust may represent the main annual summer crystal growth. A sample collected from a dry nearshore pool during July 1992 showed a thin (0.5 cm) fine white crust, resting upon the main epsomite crust. XRD and chemical analyses showed this surficial crust to contain halite (~20%), and a sodium sulfate (probably mirabilite, ~30%), in addition to epsomite (~50%). This surficial layer was not recorded during October 1987. Samples of residual brine collected from a central pool in July 1992 after an epsomite crust had formed at the surface were allowed to evaporate to complete dryness at room temperature. In addition to epsomite, XRD analyses revealed the presence of halite (NaCl), bloedite ($Na_2SO_4.MgSO_4.4H_2O$), and bishofite ($MgCl_2.6H_2O$). Although halite and bloedite are found in the saline pan, natural bishofite has not been confirmed.

From a depth of about 2 cm to about 30 cm, the epsomite shows evidence of repeated dissolution and recrystallization, with numerous irregularly undulating, corrosional surfaces and pits, and bluish black mud inclusions, composed of gypsum and Mg-carbonates (magnesite, hydromagnesite). Although recognizable near the surface, the vertical prismatic crystals are replaced in the central pool at 5 to 10 cm depth by a finer, more equant crystalline mosaic with crystals 0.5 to 2 cm across, lacking preferential crystal orientation. This layer probably represents multiple phases of dissolution and recrystallization of older summer crusts, associated with annual brine freshening.

From ~30 to 40 cm depth, the epsomite is pale yellow, tightly cemented, and is accompanied by small (10 to 15 wt.%) amounts of sodium-magnesium sulfate, probably bloedite, according to X-ray diffraction. The epsomite crystals are ~1 to 2 cm long, forming poorly defined layers, in places separated by anastomosing mud films ranging from 2 to 30 mm in thickness. Although some crystals show vague subvertical elongation, most are nonaligned mosaics.

At a depth of 40 to 50 cm in the central pool, there is a discontinuous, disrupted, bluish black mud layer, composed mostly of magnesite with subsidiary hydromagnesite, epsomite, and gypsum. This may represent a period of brine freshening when magnesium carbonates were precipitated lakewide perhaps for several years, the soluble salts crystallizing from interstitial pore waters.

Below the muddy unit is a unit of coarse, subvertically oriented, bluish black epsomite crystals up to 15 cm long, which appears to extend across the width of the pools. The crystals are rich in carbonate-gypsum mud inclusions and appear to have grown poikilotopically from interstitial brines, enclosing the host muds. Below the coarse epsomite layer, at a depth of 80 to 100 cm, are black sticky muds, composed of magnesite, hydromagnesite, and gypsum with scattered small (0.5 to 2 cm) epsomite crystals. The excavation was made at the margin of the central pool, so it is possible that thicker salt deposits and/or a different stratigraphy may occur in the pool center. Augering below this layer in the shoreline pool revealed bluish black gray muds composed mainly of fine-grained magnesite and gypsum.

The interpool muds are composed of a variable mixture of gypsum and carbonates (mostly hydromagnesite at the surface, but with magnesite and dolomite at depth), with some interstitial epsomite, fine siliciclastics, and organic matter. The siliciclastics are mainly quartz, feldspar, mica, and clay minerals, principally chlorite and illite. In summer, the pool rims have surficial efflorescent crusts, mostly of hexahydrite, bloedite, gypsum, and halite. Kieserite ($MgSO_4.H_2O$) was also recorded in several samples, but it is uncertain whether this is present in the field, or represents further dehydration of hexahydrite between collection and XRD analysis.

The evaporites in the brine pools show considerable post-depositional modification. In midsummer, tepee structures composed of epsomite plates with hexahydrite-bloedite-halite efflorescence are well developed within the larger, central pools. Mud-volcanoes, intrusive mud mounds, curvilinear mud-dikes (some containing small pebbles), and irregular sheetlike extrusions, composed mostly of blackish Mg-carbonates and gypsum, commonly form in the autumn. Liquefaction and upward injection of the underlying brine-saturated muds may result from loading of low-permeability evaporites precipitated in late summer and/or from winter ice, which is up to 25 cm thick. These processes probably account for the irregular muddy layers and clots recorded in the near-surface evaporites. Last (1989) has illustrated similar features from Ceylon Lake, Saskatchewan. Similar structures also form in the brine pools of Last Chance Lake and the Basque Lakes (Renaut and Stead, 1994a, b).

While many saline pans develop compartmentalized surfaces as a result of polygonal fracturing, tepee formation, localized spring seepages, and soft-sediment deformation and injection, the separation of brine pools by mud rims is exceptionally well developed in hypersaline lakes of the Pacific Northwest, where they are locally termed "spotted lakes" (Jenkins, 1918). Lakes exhibiting this structure include several of the Basque Lakes (Goudge, 1926a; Nesbitt, 1990; Renaut and Stead, 1994a), Spotted Lake (Jenkins, 1918; Cole, 1924), Last Chance Lake (Renaut and Long, 1989; Renaut and Stead, 1994b), several Washington lakes (Jenkins, 1918), and Alkali Lake, Oregon (Allison and Mason, 1947). Similar structures also occur in Russian hypersaline lakes (Strakhov, 1970).

Reinecke (1920) proposed that the pool structure resulted

from the sinking of bottom-nucleated salts into soft muds. Almost all occurrences of spotted lakes in the Pacific Northwest are found in steep-margined basins. Others have favored an origin by the upwelling of saline groundwater springs and seepage through lake-floor sediments, probably under artesian flow (Jenkins, 1918; Goudge, 1926a; Allison and Mason, 1947; Renaut and Long, 1989; Nesbitt, 1990; Smoot and Lowenstein, 1991). Goudge (1926a) and Nesbitt (1990) referred to similar structures at the Basque Lakes as "conical springs." At Clinton Lake the pool diameters are much larger toward the steeper northern and eastern margins from where most recharge appears to take place (Fig. 6). Indeed, the northeastern margins and north-central parts of the lake are commonly the last to dry out in summer. This suggests that saline groundwater may be upwelling through the muds that underlie the evaporites.

These structures and their hydrogeology have never been studied in detail and remain poorly understood. The temporal relationship between the interpool muds and the evaporites within the brine pools is particularly difficult to interpret. It is unclear under what conditions the carbonates and gypsum that form the interpool muds have formed. They may be early precipitates on the shallowest parts of the lake floor (including peripheral mudflats) as the lake begins to desiccate following spring flooding. However, some carbonate and gypsum might also be expected to precipitate in the brine pools. Although some buried carbonate-gypsum layers are present in the pool sequences (Fig. 13), the surficial epsomite crusts are commonly of high purity and generally gypsum-free. Close monitoring throughout the year will be required to understand fully the processes that are operative.

MINERAL ZONATION AND ORIGIN OF THE BASIN WATERS

Figure 14 summarizes the general mineral distribution in the Clinton Lake basin. As with most saline pans and playas (e.g., Hardie et al., 1978; Eugster and Hardie, 1978; Pueyo, 1978), there is a typical zonation from fringing, sparingly soluble carbonates, through moderately soluble salts to highly soluble salts. This reflects the fractionation and selective removal of ions from the waters by precipitation and other processes as they progressively evaporate to become hypersaline brines (Eugster and Hardie, 1978; Eugster, 1980). With the vegetated catchment and little overland flow, much of the early increase in salinity probably results from near-surface evapotranspiration. However, at Clinton Lake the mineral zonation does not follow a simple "bulls-eye" pattern. Although they may be in groundwater connection, Clinton Pond and Clinton Lake appear to be acting today largely as independent subbasins. If, as seems likely, groundwater is seeping from below the floor of Clinton Lake, much as at Basque Lake No. 2, then the mineralogy and its zonation are also controlled by its composition, and the loci and rate of upwelling. Although Clinton Lake is probably a hydrologically closed basin, Clinton Pond may represent an evaporation pan that periodically leaks, given the occurrence of linear hummocky ground south of the playa that suggests southward shallow flow of groundwater.

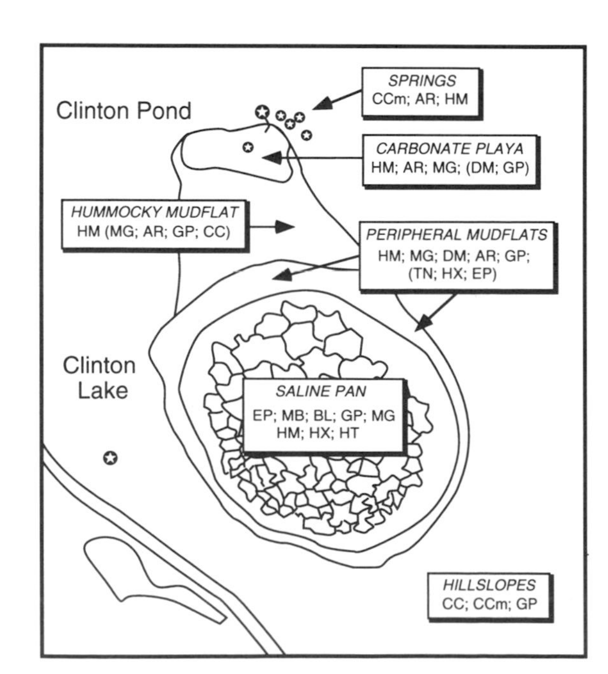

CC: calcite: $CaCO_3$
CCm: Mg-calcite (up to 17 mol% Mg)
AR: aragonite: $CaCO_3$
DM: dolomite: $CaMg(CO_3)_2$
MG: magnesite: $MgCO_3$
HM: hydromagnesite: $Mg_5(CO_3)_4(OH)_2.4H_2O$
GP: gypsum: $CaSO_4.2H_2O$
EP: epsomite: $MgSO_4.7H_2O$
HX: hexahydrite: $MgSO_4.6H_2O$
MB: mirabilite: $Na_2SO_4.10H_2O$
BL: bloedite: $Na_2SO_4.MgSO_4.4H_2O$
TN: thenardite: Na_2SO_4
HT: halite: NaCl

Figure 14. Generalized authigenic mineral distribution in the Clinton Lake basin. Minerals in parentheses are of lesser abundance.

At present, only a tentative discussion of the origin and evolution of the brines is appropriate. Seasonal variations in the composition of thc lake and groundwaters (particularly for autumn and winter) and the basin hydrogeology are largely unknown. There are significant compositional gaps in water sampling, particularly the major one between shoreline lakeward seepage (<12 g l^{-1} TDS) and the Clinton Lake brines (>250 g l^{-1} TDS). Physical conditions have prevented sampling of brines at their more dilute stages.

Figure 15 shows log-log plots for selected solutes, using chloride as a tracer. Chloride is unlikely to be removed quantitatively except in the most concentrated brines, from which halite is precipitated. Although minor chloride could be removed by deflation of halite in efflorescent crusts, these are usually only present in late summer when winds are relatively ineffective. Although the validity of using chloride as a conservative tracer remains to be proven, it is nonetheless useful in illustrating general aspects of the basin-water behavior. Several analyses of Alkali Valley saline lakes (Table 1) are included on the log plots for comparison because their waters have evolved on the same bedrock as Clinton Lake.

The probable origins of the major ions are discussed in Renaut (1990). The low Na/Cl ratios in dilute runoff are typical for local waters derived from the Cache Creek limestones, argillites, and cherts. The high sulfate concentration in runoff may be explained by oxidation of sulfides (mainly pyrite) in the Cache Creek shales and argillites, and the dissolution of gypsum, which is locally abundant along shear zones (Cole, 1913; Renaut, unpublished).

From Table 1 and Figure 15, it is clear that the springs and dilute seepage waters cannot evolve by simple evaporative concentration of percolated runoff waters with the sampled compositions if both Na^+ and Cl^- are behaving conservatively. Between runoff and the spring waters the molar Na/Cl ratio has increased on average approximately threefold. Mg^{2+}, Na^+, K^+, SO_4^{2-}, and CO_2 species have been acquired in the subsurface by circulating groundwaters in contact with the Cache Creek rocks and surficial sediments. The magnesium enrichment is exceptionally high, reflecting in part the weathering and alteration of Mg-silicates. Mafic and ultramafic rocks, including serpentinites, greenstones, and basalts, are locally very common.

Initially, calcium is the principal cation in dilute runoff, but most is removed by early precipitation of micritic calcite and Mg-calcite cements, and minor gypsum, in soils and till (Renaut, 1990). The emergent spring waters north and east of Clinton Pond have molar Mg/Ca ratios from 22 to >60, partly as a consequence of calcite and gypsum precipitation, but mainly due contact with mafic bedrock. While the chloride concentration has increased by a factor of 3 to 4, the Mg concentration in the springs has increased to at least 15 times that in runoff.

Calcite, aragonite, and hydromagnesite are precipitated around the springs, by a combination of evaporative concentration, degassing of CO_2, and possibly with some microbial mediation. Most remaining spring-derived Ca^{2+} is then consumed in Clinton Pond either as aragonite or dolomite, or by small amounts of gypsum in efflorescent crusts. The Mg-enriched waters, with a Mg/Ca ratio of >150, then precipitate hydromagnesite, further depleting the playa waters of much of their remaining carbonate and bicarbonate. Clinton Pond waters in summer are supersaturated with respect to hydromagnesite by a factor of 1.7 to 3, as well as dolomite and aragonite. With further precipitation of carbonate species, the residual waters gradually evolve to become Mg-Na-SO_4 in composition.

At present Clinton Pond is a topographically closed basin. The fate of any residual groundwater is unknown, but because the playa floor lies about a meter above Clinton Lake and the carbonates are underlain by sands and gravels, some southward seepage is feasible. Residual waters, both from Clinton Pond and the springs that bypass the playa, may flow toward Clinton Lake through the shallow permeable siliciclastic sediments that underlie the carbonates of the hummocky mudflats. The aligned hummocks of hydromagnesite and associates soil heave observed between Clinton Pond and the northeast corner of Clinton Lake (Fig. 11), may be evidence for shallow southward movement of groundwater. Whether this is presently active is unknown.

Four relatively dilute (1 to 2 g l^{-1} TDS) seepage waters (BCS-10 and 12, CB-13 and 14 in Table 1) were sampled from shallow pits (50 to 100 cm deep) dug into the narrow mudflat surrounding Clinton Lake. Given their relatively low salinity, these analyses are taken to represent samples of very shallow (unconfined?) groundwater flowing into the saline pan, but which may have mixed variably with more evolved brines from the pan itself. Based on the molar Na/Cl ratios of the groundwaters, two types of inflow are recognized—one having molar Na/Cl and Na/K ratios similar to the springs near Clinton Pond (CB-13 and 14), and one (BCS-12) with much higher molar Na/Cl and Na/K ratios, closer in composition to the waters of Three Mile Lake. Sample BCS-10 from the northeastern shore has an intermediate composition, with a Na/K ratio similar to the spring seeps northeast of Clinton Lake, but with a higher Na/Cl ratio, suggesting waters of mixed derivation. This limited evidence suggests that Clinton Lake is probably fed by groundwaters with at least two different compositions—one similar to the springs feeding Clinton Pond, and a second source, possibly similar to the aquifer that feeds Three Mile Lake.

The source of the waters with high Na/Cl ratios (>20) is unclear. They may derive from dilute waters with an initially higher Na/Cl ratio than locally sampled slope runoff. Although uncommon, dilute (0.3 g l^{-1} TDS) runoff waters with molar Na/Cl ratios exceeding 20 have been found on till-covered basalts of the Cariboo Plateau (Renaut, 1990). A second possibility is that they derive from a common aquifer with the springs, but have preferentially acquired more Na^+ at an intermediate stage in their evolution, possibly by cation exchange. De Deckker and Last (1989) proposed the reaction: 2Na-clay

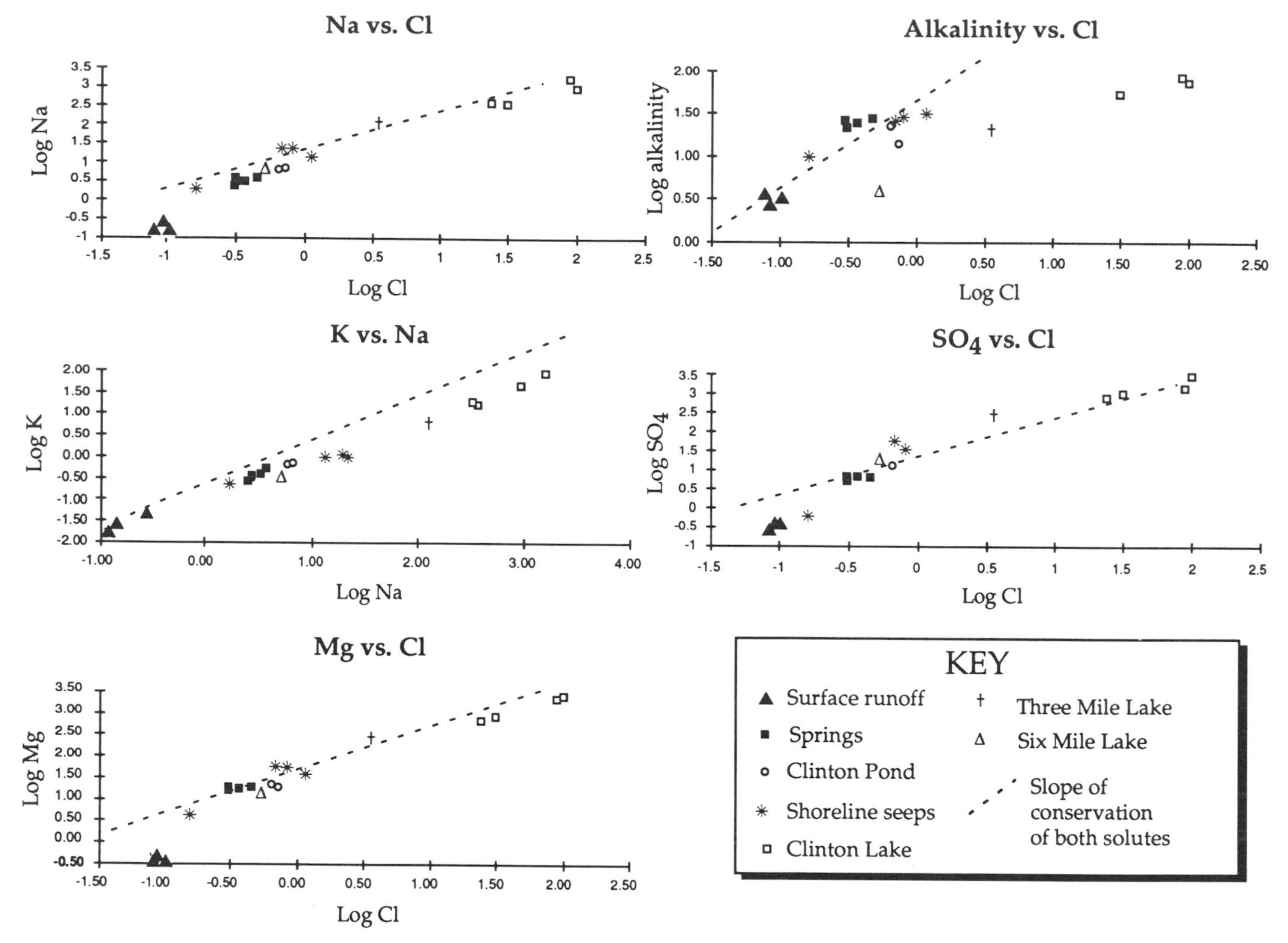

Figure 15. Log-log plots (molar basis) for selected solutes versus Cl in the Clinton basin.

+ Ca^{2+} → Ca-clay + $2Na^+$ as a method to explain loss of Ca^{2+} from groundwaters in Australian evaporitic playas. However, this process could also increase the Na/Cl ratio in shallow groundwaters. A third possibility is that the seepage waters with a Na/Cl ratio similar to the springs have mixed with lake marginal brines having a higher Na/Cl ratio. In support, sample BCS-12 is moderately saline (12 g l^{-1} TDS) and is enriched in sulfate. However, the Na/Cl ratios of the analyzed lake brine are *lower* than BCS-12, making it difficult to explain by mixing alone. Alternatively, the seeps with a high Na/Cl ratio may result from mixing of dilute seepage or runoff waters with lake brines having a higher Na/Cl ratio than those that existed when sampled. Significant diurnal changes in the Na/Cl ratio of the brine would result, for example, by ephemeral mirabilite crystallization.

At present, there is no clear explanation for the derivation of the seepage waters with the high Na/Cl ratios. The simplest, and most likely, explanation is that they derive from a groundwater source that has not been sampled at its more dilute stages. In support, Three Mile Lake (BC-14 in Table 1), which has very similar Na/Cl and Na/K ratios to the southwestern shoreline seepage, derives most of its water from direct groundwater discharge. Several other lakes in Alkali Valley also have comparable Na/K ratios. The most dilute water sample reported by Nesbitt (1990) at Basque Lake No. 2, with a catchment entirely in similar Cache Creek lithology, also has a high molar Na/Cl ratio (~19). Longer residence time in the aquifer, slightly deeper circulation, local lithological variations, strong seasonal variations in the Na/Cl ratio in the lake brine, or another, as yet undiscovered mechanism may account for the variable composition of the lake marginal seepage.

The contrasting compositions of the shoreline seeps and the lake brine suggest that Clinton Lake is being fed by some moderately fresh, probably very shallow, unconfined groundwaters, but also possibly by more evolved saline groundwaters of slightly deeper (?artesian) and/or longer circulation. As discussed, the basin hydrogeology is very poorly known. Elsewhere in the region, glaciofluvial sands and gravels are thought to be important aquifers (Halstead, 1991; Renaut, 1993), and these are likely to be present below the surficial sediments of Alkali Valley. Although the brine pools may be in contact with such an aquifer, it is difficult to reconcile the preserved stratigraphy of the pools (Fig. 13) and the occurrence of highly soluble salts such as bloedite with a *major* upwelling of groundwater unless that water is strongly saline. Dilute waters would likely lead to dissolution of soluble salts and a collapsed stratigraphy unless the orifice at the base of the pool is very narrow or the flow rate relatively low. Goudge (1926a) recorded saline "springs" with a salinity of >300 g l^{-1} TDS flowing from the margins and narrow openings at the base of pools at Basque Lake No. 1. Their origin was not explained, but they must be products of evaporative concentration of waters within the very small catchment, probably under artesian pressures. Last (1989) showed that large quantities of salts and ice, including mounds 50 cm high, are formed around active sublacustrine springs at Ceylon Lake, Saskatchewan. These clearly differ from those at Clinton Lake, suggesting more passive or less focused discharge of groundwater.

The dilute seepage waters feeding Clinton Lake explain the presence of Ca-bearing carbonates and gypsum in the peripheral mudflats and in the interpool muds. The zone with aragonite and dolomite is rapidly succeeded by a zone of hydromagnesite and magnesite, but this is not a simple concentric pattern; the different loci of dilute seepage appear to control the spatial distribution of the carbonates.

The processes that control brine composition and the sequence of evaporite mineral precipitation within Clinton Lake remain poorly understood. After precipitation of hydromagnesite and/or magnesite, an Mg-Na-SO_4 brine is produced, from which epsomite is the main precipitate. Remaining calcium derived from dilute surface inflow, seepage, or upwelling groundwater is removed mainly as aragonite, gypsum, or possibly disordered dolomite. As at the Basque Lake No. 2 (Nesbitt, 1990), it is unclear at which stage(s) gypsum is precipitated. Crystallization of magnesium sulfate as epsomite or ephemeral hexahydrite crusts, relatively enriches the brine in sodium. Some of the sodium is precipitated as bloedite. The mineral appears to be less abundant than in the Basque Lakes, but was recorded with epsomite in the 1992 surface crusts, and is present in the underlying salts and muds. It is unclear whether it is precipitated subaqueously by evaporation in the pools or whether it is an interstitial or diagenetic mineral after epsomite (Ordóñez et al., 1994). Sodium is also removed as halite on desiccation of the brine pools in late summer, and as thin mirabilite crusts in spring, autumn, and winter on epsomite. Unless protected by a skin of detrital mud or precipitated hydromagnesite, most of the sodium-bearing salts are returned to the brine when the temperature rises in spring, increasing the Na/Cl ratio (Table 1). Some sulfate is being lost by bacterial reduction in the lake muds and peripheral mudflats as demonstrated by a strong odor of H_2S when the muds are disturbed. Some of the bicarbonate in Clinton Lake brine may be produced from organic matter, associated with anoxic sulfate reduction in the lake muds.

Silica is depleted at almost all stages relative to chloride, but the sinks are unknown. Pennate diatoms are found in microbial mats around the springs and in surficial aragonite muds of Clinton Pond, accounting for the silica depletion between the springs and pond waters (Table 1). However, only rare corroded diatoms are found in the underlying muds, demonstrating early dissolution in the alkaline pore waters. The fate of this silica is unknown, but sepiolite, palygorskite, and opal-A are found in association with Mg-carbonates elsewhere in the Cariboo region (Renaut, 1993). The low silica in the groundwaters and the saline pan brine may also be due to sorption or neoformed clay minerals, but their mineralogy has not yet been investigated. Potassium appears to be conserved throughout most stages, but some is removed from the most concentrated brines, either as K-bearing silicates or adsorption by clay minerals (cf. Nesbitt, 1990). Potash salts have not been recorded. Figure 16 summarizes the inferred origin of the Clinton Lake brine.

HISTORICAL DEVELOPMENT OF CLINTON LAKE

When he visited Clinton in 1918 or 1919, Reinecke (1920, p. 53) was informed by older residents of the village that Clinton Lake was "once an irrigated hay meadow, periodically occupied by a lake." They also informed him that, "the deposits of epsomite have been in evidence for the last few years only." It remains to be demonstrated whether the evaporites are as young as he implies, although irrigation may have destroyed any preexisting salts. If, as Reinecke suggests, epsomite precipitation only began last century, then an explanation must be sought. One possibility is that construction of the

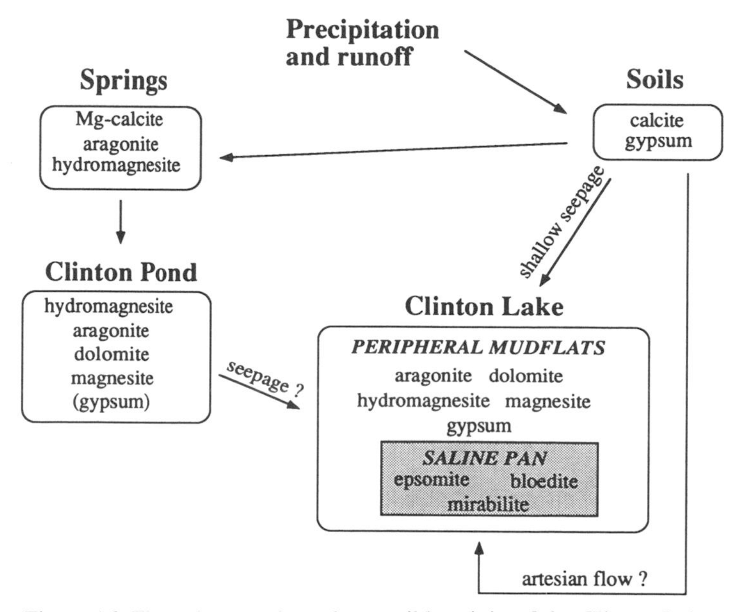

Figure 16. Flow chart to show the possible origin of the Clinton Lake brines.

Cariboo Wagon Road during the Cariboo gold-rush of 1860s (Waite, 1988) may have modified the drainage basin. The road, now British Columbia Highway 97, follows a rise in the ground west of the lake that at least in places has been artificially embanked. Several groundwater-fed saline ponds directly west of the highway (Fig. 6) are apparently ponded by the embankment. Some of their recharge waters may originally have fed the Clinton Lake catchment. However, because it lies in a natural hydrographic sump and is mainly groundwater fed, Clinton Lake is likely to have been present before any artificial modification.

The area north of Clinton Pond shows evidence of artificial disturbance and a trench was dug in 1975 near the springs to provide water for grazing animals. The effects of these or earlier modifications on the natural system are unclear. Although Reinecke did not show the playa on his map (Reinecke, 1920, Fig. 8), the thickness (locally >70 cm) and lateral extent of the carbonates suggest that a ponded area of damp ground likely predated any man-made modifications.

Northwest of Clinton Lake there is a group of small lakes and ponds, separated by sinuous ridges. All are groundwater-fed, there being no channelled inflow. The largest is precipitating hydromagnesite, which takes the form of small (<1 cm) granules. At some stage(s) during the Holocene these lakes may have been connected to a fresher Clinton Lake, but aerial photographs suggest that they have been separate topographically closed basins for a considerable time.

At present, the deposits have not been dated, nor have their sedimentation rates been assessed. The Bridge River tephra, dated at ~2,350 BP (Mathewes and Westgate, 1980), is exposed in colluvial deposits in Alkali Valley (e.g., in the roadcut at Six Mile Lake), but has not yet been found in or below the lake sediments. Aerial photographs show slight changes in vegetation that suggest former higher shorelines, but most are animal tracks or appear unconvincing as paleoshorelines on the ground. However, the presence of Mg-carbonates containing rootlets, both in the inner mudflat sediments and below the evaporites, is clear evidence for a former, somewhat fresher, alkaline lake. Many valleys in the region, including parts of Cutoff Valley east and west of Clinton, are floored by damp, grass-covered meadows composed predominantly of massive hydromagnesite and magnesite muds (Reinecke, 1920; Cummings, 1940; Grant, 1987; Renaut, unpublished). Most show hummocky ground suggestive of sediments formed displacively or modified by groundwater. About 1 km southeast of Clinton, fossil hydromagnesite deposits are undergoing incision, providing evidence for former regionally higher water tables. Local travertine deposits also provide evidence for former mid-Holocene moister conditions (Renaut and Long, 1986).

CONCLUSIONS

Although epsomite is a common accessory evaporite mineral in many sodium sulfate lakes (e.g., Last, 1984, 1989), few modern magnesium sulfate lakes with abundant epsomite deposits have been described, the exceptions being the Basque Lakes (Nesbitt, 1974, 1990; Eugster and Hardie, 1978) and several Spanish saline lakes (Pueyo and De la Peña, 1991). Clinton Lake provides a further rare example of a modern Mg-Na-SO_4 saline pan.

Although similar to the Basque Lakes in morphology, mineralogy, and hydrochemistry, Clinton Lake basin shows a wider range of depositional subenvironments and may have a more complex hydrogeological setting. Preliminary interpretation of the brine evolution suggests that the waters probably follow the same general path as the Basque Lake No. 2 (Path $IIIB_{1b}$ of Eugster and Hardie, 1978), except that hydromagnesite is an important alkaline earth carbonate at Clinton Lake. Hydromagnesite deposits similar to those at Clinton Pond occur at Basque Lake No. 5, a shallow evaporative playa (Renaut and Stead, 1993a), but the mineral was not reported at Lake No. 2.

Early precipitation of Ca-bearing carbonates (calcite, aragonite) and gypsum in soils, and at springs and seepage sites, raises the Mg/Ca ratio in the waters. Aragonite and hydromagnesite then precipitate both subaqueously (Clinton Pond) and from shallow groundwater (around springs and in mudflats). Although not confirmed, these minerals may then undergo early diagenetic alteration to dolomite and magnesite, respectively. The dolomite occurs in sulfate-reduced muds in mudflats around Clinton Lake and in surficial muds at Clinton Pond. Gypsum precipitation in Clinton Lake removes remaining calcium, and magnesite most remaining carbonate, leaving a Mg-Na-SO_4 brine. With evaporative concentration epsomite crystallizes annually as a clear crust in the mud-rimmed brine pools that form the floor of Clinton Lake. Mirabilite crystallizes during periods of brine cooling, but most redissolves upon warming. Bloedite is found in the surface crusts, shallow subsurface evaporites, and in efflorescence, but its genesis is poorly understood. The sediments from the lake floor provide evidence for multiple phases of dissolution and reprecipitation of evaporites, together with extensive physical disruption that is probably associated with gravitational loading by salt and ice, and displacive salt growth. Carbonates underlying and interlayered with the evaporites provide evidence for somewhat fresher Holocene precursor lakes. Evidence acquired to date suggests that the lake brines are replenished mostly by groundwater. Some dilute waters seep into the lake through the surrounding permeable surficial sediments. Other waters, but probably of somewhat higher salinity, may be upwelling under shallow artesian flow through narrow orifices in the brine pools that form the floor of the saline pan.

Further detailed analyses of the hydrogeology of the basin, temporal variations in hydrochemistry and evaporite precipitation, and chemistry of the carbonate minerals are required to understand fully the origin and evolution of the Clinton Lake brines.

ACKNOWLEDGMENTS

This work was supported by grants from the Natural Sciences and Engineering Research Council of Canada and the British Columbia Geoscience Research Grant Program, which are gratefully acknowledged. I thank Doug Stead, Peter Long, Tony Gonzales, Bill Last, and Geoff Koehler for assistance with fieldwork. Access to the land was kindly granted by Mr. A. Danielson. I also thank Drs. John Warren and Michael Rosen for their comments and ideas that helped to improve the original manuscript.

REFERENCES CITED

Allison, I. S., and Mason, R. S., 1947, Sodium salts of Lake County, Oregon: Oregon Department of Geology and Mineral Industries Short Paper 17, 12 p.

Annas, R., and Coupe, D., 1979, Biogeoclimatic zones and subzones of the Cariboo Forest Region: British Columbia Ministry of Forests, Victoria, 99 p.

Atmospheric Environment Service, 1982, Canadian climate normals 1951–1980: Temperature and precipitation (British Columbia): Vancouver, Environment Canada, 268 p.

Ball, J. W., Nordstrom, D. K., and Zachmann, D. W., 1987, WATEQ4F—A personal computer Fortran translation of the geochemical model WATEQ2 with revised data base: U.S. Geological Survey Open-File Report 87-50, 63 p.

Campbell, R. B., and Tipper, H. W., 1971, Geology of the Bonaparte Lake map area, British Columbia: Geological Survey of Canada Memoir 363, 100 p.

Chafetz, H. S., and Folk, R. L., 1984, Travertines: Depositional morphology and bacterially constructed constituents: Journal of Sedimentary Petrology, v. 54, p. 289–317.

Clague, J. J., 1981, Late Quaternary geology and geochronology of British Columbia, Part 2: Summary and discussion of radiocarbon dated Quaternary history: Geological Survey of Canada Paper 80-35, 41 p.

Clague, J. J., and 7 others, 1989, Quaternary geology of the Canadian Cordillera, *in* Fulton, R. J., ed., Quaternary geology of Canada and Greenland: Geological Survey of Canada, Geology of Canada, no. 1, p. 15–96.

Cole, L. H., 1913, Gypsum in Canada: Canada Department of Mines, Mines Branch Report, 245, 256 p.

Cole, L. H., 1924, Sodium and magnesium salts of western Canada: Transactions of the Canadian Institute of Mining and Metallurgy, v. 27, p. 209–247.

Cummings, J. M., 1940, Saline and hydromagnesite deposits of British Columbia: British Columbia Department of Mines Bulletin 4, 160 p.

De Deckker, P., and Last, W. M., 1989, Modern, non-marine dolomite in evaporitic playas of western Victoria, Australia: Sedimentary Geology, v. 64, p. 223–238.

Eugster, H. P., 1980, Geochemistry of evaporitic lacustrine deposits: Annual Reviews in Earth and Planetary Sciences, v. 8, p. 35–63.

Eugster, H. P., and Hardie, L. A., 1978, Saline lakes, *in* Lerman, A., ed., Lakes: Chemistry, geology, physics: New York, Springer, p. 237–293.

Fulton, R. J., 1984, Quaternary glaciation, Canadian Cordillera, *in* Fulton, R. J., ed., Quaternary stratigraphy of Canada—A Canadian contribution to I.G.C.P. Project 24: Geological Survey of Canada Paper 84-10, p. 39–47.

Goudge, M. F., 1926a, Magnesium sulphate in British Columbia: Canada Department of Mines, Mines Branch Report, 642, p. 62–80.

Goudge, M. F., 1926b, Sodium carbonate in British Columbia: Canada Department of Mines, Mines Branch Report, 642, p. 81–102.

Grant, B., 1987, Magnesite, brucite and hydromagnesite occurrences in British Columbia: British Columbia Geological Survey Branch Open-File Report 1987-13, 68 p.

Halstead, E. C., 1991, Energy and groundwater resources of the Canadian Cordillera: Part E: Groundwater, *in* Gabrielse, H., and Yorath, C. J., eds., Geology of the Cordilleran orogen in Canada: Geological Survey of Canada, Geology of Canada, no. 4, p. 793–801.

Hammer, U. T., and Forró, L., 1992, Zooplankton distribution and abundance in saline lakes of British Columbia, Canada: International Journal of Salt Lake Research, v. 1, p. 65–80.

Hardie, L. A., 1968, The origin of the Recent non-marine evaporite deposit of Saline Valley, Inyo County, California: Geochimica et Cosmochimica Acta, v. 32, p. 1279–1301.

Hardie, L. A., Smoot, J. P., and Eugster, H. P., 1978, Saline lakes and their deposits: A sedimentological approach, *in* Matter, A., and Tucker, M. E., eds., Modern and ancient lake sediments: Oxford, Blackwell, p. 7–41.

Heimann, A., and Sass, E., 1989, Travertines in the northern Hula Valley, Israel: Sedimentology, v. 36, p. 95–108.

Jenkins, O. P., 1918, Spotted lakes of epsomite in Washington and British Columbia: American Journal of Science, v. 46, p. 638–644.

Julia, R., 1983, Travertines, *in* Scholle, P. A., Bebout, D. G., and Moore, C. H., eds., Carbonate depositional environments: American Association of Petroleum Geologists Memoir 33, p. 64–73.

Last, W. M., 1984, Sedimentology of playa lakes of the northern Great Plains: Canadian Journal of Earth Sciences, v. 21, p. 107–125.

Last, W. M., 1989, Sedimentology of a saline playa in the northern Great Plains, Canada: Sedimentology, v. 36, p. 109–123.

Lowenstein, T., and Hardie, L. A., 1985, Criteria for recognition of salt-pan evaporites: Sedimentology, v. 32, p. 627–644.

Mathewes, R. W., and Westgate, J. A., 1980, Bridge River tephra: Revised distribution and significance for detecting old carbon errors in radiocarbon dates of limnic sediments in southern British Columbia: Canadian Journal of Earth Sciences, v. 17, p. 1454–1461.

Monger, J.W.H., 1989, Overview of Cordilleran geology, *in* Rickets, B. D., ed., Western Canada sedimentary basin—A case history: Calgary, Canadian Society of Petroleum Geologists, p. 9–32.

Monger, J.W.H., and 9 others, 1991, Upper Devonian to Middle Jurassic Assemblages, Part B: Cordilleran terranes, *in* Gabrielse, H. and Yorath, C. J., eds., Geology of the Cordilleran orogen in Canada: Geological Survey of Canada, Geology of Canada, no. 4, p. 281–327.

Nesbitt, H. W., 1974, The study of some mineral-aqueous solution interactions [Ph.D. thesis]: Johns Hopkins University, 173 p.

Nesbitt, H. W., 1990, Groundwater evolution, authigenic carbonates and sulphates, of the Basque Lake No. 2 Basin, Canada, *in* Spencer, R. J., and Chou, I-Ming, eds., Fluid-mineral interactions: A tribute to H. P. Eugster: Geochemical Society Special Publication No. 2, p. 355–371.

Ordóñez, S., Sánchez Moral, S., García Del Cura, M. A., and Rodríguez Badiola, E., 1994, Precipitation of salts from Mg-(Na)-SO_4-Cl playa-lake brines: The endorheic saline ponds of La Mancha, central Spain, *in* Renaut, R. W., and Last, W. M., eds., Sedimentology and geochemistry of modern and ancient saline lakes: Society for Sedimentary Geology (SEPM) Special Publication, (in press).

Pueyo, J. J., 1978/79, La precipitación evaporitica actual en las lagunas saladas del área: Bujaraloz, Sástago, Caspe, Alcañiz y Calanda (provincias de Zaragoza y Teruel): Revista del Instituto de Investigaciones Geológicas Diputación Provincial Universitat de Barcelona, v. 33, p. 5–56.

Pueyo, J. J., and De la Peña, J. A., 1991, Los lagos salinos Españoles: Sedimentología, hidroquímica y diagénesis, *in* Pueyo, J. J., ed., Génesis de formaciones evaporîticas: Modelos Andinos e Ibéricos: Publicacions de la Universitat de Barcelona, Estudi General, p. 163–192.

Reinecke, L., 1920, Mineral deposits between Lillooet and Prince George, British Columbia: Geological Survey of Canada Memoir 118, 129 p.

Renaut, R. W., 1990, Recent carbonate sedimentation and brine evolution in the saline lake basins of the Cariboo Plateau, British Columbia, Canada:

Hydrobiologia, v. 197, p. 67–81.

Renaut, R. W., 1993, Morphology, distribution, and preservation potential of microbial mats in the hydromagnesite-magnesite playas of the Cariboo Plateau, British Columbia, Canada: Hydrobiologia, v. 267, p. 75–98.

Renaut, R. W., and Long, P. R., 1986, Post-glacial travertine deposits of the Clinton area, Interior British Columbia [abs.]: Geological Association of Canada and Mineralogical Association of Canada, Joint Annual Meeting, Programme with Abstracts, v. 11, p. 117.

Renaut, R. W., and Long, P. R., 1987, Freeze-out precipitation of salts in saline lakes—Examples from western Canada, *in* Strathdee, G. L., Klein, M. O., and Melis, L. A., eds., Crystallization and precipitation: Oxford, Pergamon, p. 33–42.

Renaut, R. W., and Long, P. R., 1989, Sedimentology of the saline lakes of the Cariboo Plateau, Interior British Columbia: Sedimentary Geology, v. 64, p. 239–264.

Renaut, R. W., and Stead, D., 1991, Recent magnesite-hydromagnesite sedimentation in playa basins of the Cariboo Plateau, British Columbia: British Columbia Geological Survey Branch Paper 1991-1, p. 279–288.

Renaut, R. W., and Stead, D., 1994a, The saline lakes of southern British Columbia, *in* Gierlowski-Kordesch, E., and Kelts, K., eds., Global geological record of lake basins, Vol. 1: Cambridge, Cambridge University Press (in press).

Renaut, R. W., and Stead, D., 1994b, Last Chance Lake—A natric playa-lake in Interior British Columbia, Canada, *in* Gierlowski-Kordesch, E., and Kelts, K., eds., Global geological record of lake basins, Vol. 1: Cambridge, Cambridge University Press (in press).

Renaut, R. W., Stead, D., and Owen, R. B., 1994, The saline lakes of the Fraser Plateau, British Columbia, Canada, *in* Gierlowski-Kordesch, E., and Kelts, K., eds., Global geological record of lake basins, Vol. 1: Cambridge, Cambridge University Press (in press).

Riding, R., 1991, Classification of microbial carbonates, *in* Riding, R., ed., Calcareous algae and stromatolites: Berlin, Springer, p. 21–51.

Rollins, L., 1989, PCWATEQ: A simple, interactive PC version of the water chemistry analysis program WATEQ4F: Woodland, California, ShadoWare.

Smoot, J. P., and Lowenstein, T. K., 1991, Depositional environments of nonmarine evaporites, *in* Melvin, J. L., ed., Evaporites, petroleum and mineral resources: Amsterdam, Elsevier, p. 189–347.

Strakhov, N. M., 1970, Principles of lithogenesis, 3: New York, Plenum, 577 p.

Topping, M. S., and Scudder, C.G.E., 1977, Some physical and chemical features of saline lakes in central British Columbia: Syesis, v. 10, p. 145–166.

Valentine, K.W.G., and Schori, A., 1980, Soils of the Lac La Hache–Clinton area, British Columbia: British Columbia Soil Survey Report 25, 118 p.

Waite, D. E., 1988, The Cariboo Gold Rush story: Surrey, British Columbia, Hancock House, 112 p.

Warren, J. K., 1982, The hydrological significance of Holocene tepees, stromatolites, and boxwork limestones in coastal salinas in south Australia: Journal of Sedimentary Petrology, v. 52, p. 1171–1201.

Warren, J. K., 1985, On the significance of evaporite lamination, in Schreiber, B. C., and Harner, H. C., eds., Sixth Symposium on Salt, 1: Alexandria, Virginia, The Salt Institute, p. 161–170.

Manuscript Accepted by the Society July 2, 1993

Printed in U.S.A.

Geological Society of America
Special Paper 289
1994

Paleohydrology of playas in the northern Great Plains: Perspectives from Palliser's Triangle

William M. Last
Department of Geological Sciences, University of Manitoba, Winnipeg, Manitoba, R3T 2N2 Canada

ABSTRACT

Palliser's Triangle, the most arid portion of the Great Plains of western Canada, contains many playa lake basins. Because of the great diversity in basin types, brine chemistries, and depositional processes, the sediments in these lakes offer a tremendous opportunity to examine past hydrological and environmental conditions and changes in the region. Despite the sensitivity of these deposits to environmental change, interpreting the records in terms of paleohydrology, chemistry, and climate is fraught with difficulty. Factors that complicate these interpretations include: diagenesis of the evaporites, post-depositional physical disruption of the sediments, and a lack of proper understanding of the depositional processes operating in lakes of this type. Furthermore, an active and growing industrial minerals industry based on the deposits of the salt playas has obliterated, and will likely continue to adversely affect, the stratigraphic records of some of the basins with the greatest research potential. Notwithstanding these problems, the sediments of the playa lakes provide the best and, in some cases, only record of past environmental conditions in this semiarid region. Paleohydrology in this area of the northern Great Plains is poised for a rapid expansion, fueled by the combination of significant technological breakthroughs, improvements in methodology, and a more positive view of the importance of paleohydrological research in environmental management.

INTRODUCTION

The northern Great Plains of western Canada form a unique setting for millions of lakes. Because of the relatively high evaporation to precipitation ratios in this region, and the presence of extensive areas of closed drainage, saline and hypersaline waters dominate these lakes. Most of the lakes that occur in Palliser's Triangle (Fig. 1), the most arid portion of this large geographic region, are shallow and exhibit playa characteristics: filling with water during the spring and early summer and drying completely by late summer/autumn.

The past decade has witnessed considerable growth in interest and research on both the modern sedimentary processes and the Holocene stratigraphic records in these saline playa basins. Indeed, it is now recognized that the playas of Palliser's Triangle provide nearly the only source of detailed, high resolution, physical and chemical paleoenvironmental information for the Holocene of the region. Unfortunately, the very aspects that make these salt-lake sediments so potentially attractive for paleohydrological, paleolimnological, and paleoclimatic analyses also give rise to significant interpretative problems.

The objectives of this paper are threefold: (1) introduce the features of the northern Great Plains, and specifically Palliser's Triangle, that are important to the occurrence of the salt lakes, briefly summarizing the main sedimentological and geochemical characteristics of the playas; (2) discuss the more serious difficulties that must be faced in attempting to interpret the Holocene stratigraphic records of these basins; and (3) highlight future paleohydrological research directions and opportunities in this region. This is not intended to be either a review or synthesis of specific playa basins in the Great Plains; several other recent publications have already done this (Last and Slezak,

Last, W. M., 1994, Paleohydrology of playas in the northern Great Plains: Perspectives from Palliser's Triangle, *in* Rosen, M. R., ed., Paleoclimate and Basin Evolution of Playa Systems: Boulder, Colorado, Geological Society of America Special Paper 289.

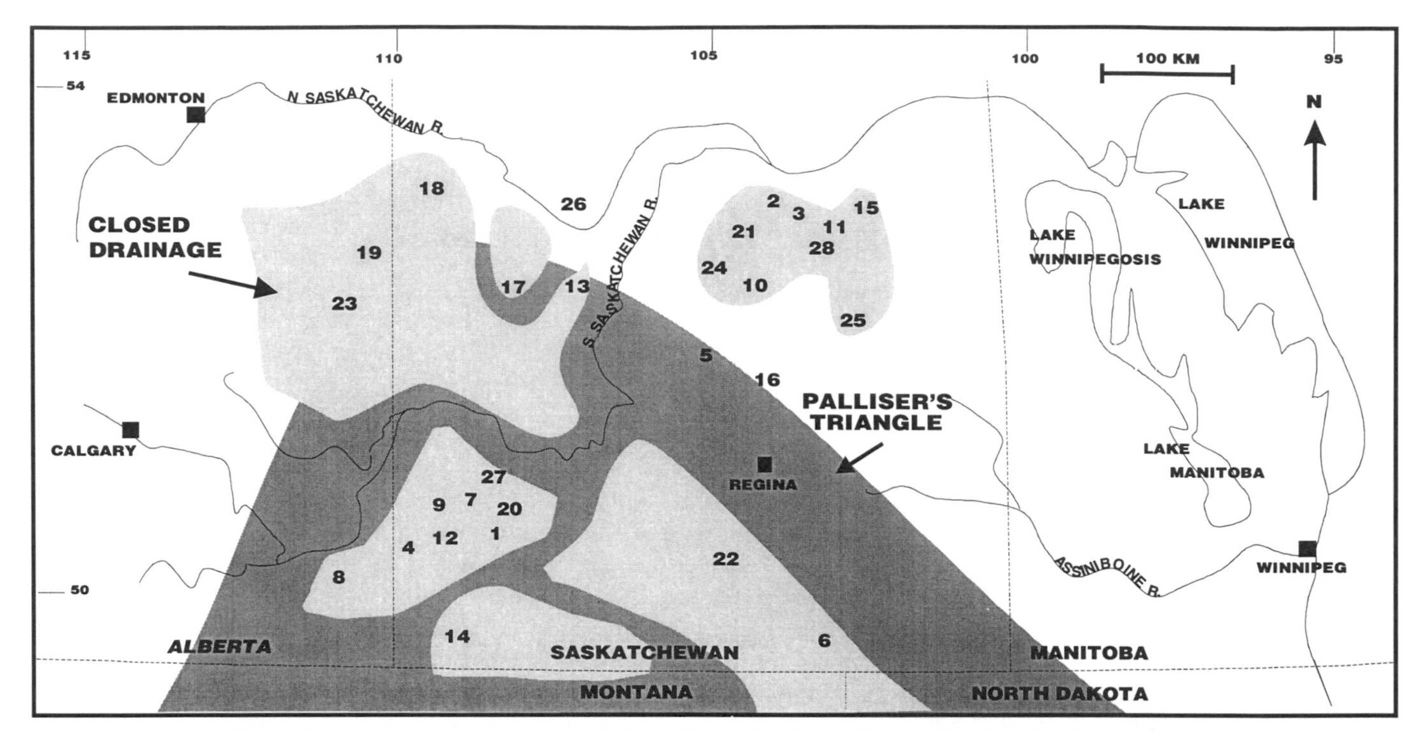

Figure 1. Location map of the northern Great Plains showing Palliser's Triangle. Also shown are the areas of internal (endoreic) drainage and the approximate locations of the lakes mentioned and discussed in the text: 1, Antelope; 2, Arthur; 3, Basin; 4, Bitter; 5, Blackstrap; 6, Ceylon; 7, Chain; 8, Chappice; 9, Corral; 10, Dana; 11, Deadmoose; 12, Freefight; 13, Grandora; 14, Harris; 15, Lenore; 16, Little Manitou; 17, Lydden; 18, Manitou; 19, Metiskow; 20, Mud; 21, Muskiki; 22, Old Wives; 23, Oliver; 24, Porter; 25, Quill; 26, Redberry; 27, Vincent; 28, Waldsea.

1988; Teller and Last, 1990). Finally, the emphasis of this paper is nonbiological. Smol's (1989) discussion on recent advances and future developments of biological paleolimnology are directly applicable to the study of the paleohydrology of playas in Palliser's Triangle. In general, however, the wealth of paleobiological information contained in the sediments of these lakes has not been as extensively exploited as elsewhere in the world (e.g., Löffler, 1987; Meriläinen et al., 1983).

WHAT DO WE KNOW ABOUT THE PLAYAS OF THE REGION?

Location and setting

. . . . in the central part of the continent there is a region, desert, or semi-desert in character, which can never be expected to become occupied by settlers . . . . Although there are fertile spots throughout its extent, it can never be of much advantage to us as a possession. (Palliser, 1862)

The northern Great Plains of Canada stretch from the Precambrian Shield near Winnipeg, Manitoba, westward for nearly 1,600 km to the Foothills of the Rocky Mountains. This is a vast region of flat to gently rolling terrain, interrupted only by occasional steep-sided and deeply entrenched river valleys. Palliser's Triangle, named after Captain John Palliser, the leader of one of the first scientific expeditions into the western interior of Canada, is informally defined as the area between longitude 100° and 114°, extending north from the 49th parallel to an apex at about 52° (Fig. 1). Although it is misleading to generalize about such a large area, Palliser's Triangle does experience a cold, semiarid steppe climate, with a mean annual temperature of about 3°C and average precipitation of 325 mm per year (CNC/IHD, 1978). The area of Palliser's Triangle is somewhat warmer and drier than the rest of the northern Great Plains. Warm summer temperatures combine with low humidity and strong winds to give the area an annual moisture deficit of generally greater than about 1 m.

Another important regional characteristic is the occurrence of large areas of internal drainage. In total, the drainage in well over half of the entire Palliser's Triangle is closed; southern Saskatchewan alone has more than 90,000 km^2 of internal drainage, representing about 10% of North America's total (Fig. 1). Although not all playas of the region occur in these endoreic areas, and not all lakes in these large areas of internal drainage are saline, the great expanse of closed drainage does play a pivotal role in development of the tremendous number of saline water bodies that occur.

Finally, the many salt lakes of the region are inextricably

linked to the geological history of the northern Great Plains. Pleistocene continental glaciation has resulted in a thick sequence of unconsolidated glacial, glaciofluvial, and glaciolacustrine sediment mantling the generally flat-lying Phanerozoic sedimentary bedrock. The origin of nearly all of the natural lake basins in the Great Plains is a direct result of this Pleistocene glaciation or of fluvial activity associated with meltwater from these glaciers.

Similarly, the dissolved ions that make the lake waters so salty are, to a major degree, ultimately derived from the thick section of Phanerozoic sedimentary bedrock and glacial deposits. The bedrock of the region consists of a sequence of Mesozoic and Cenozoic siliciclastic rocks overlying a series of Paleozoic carbonates and evaporites. Many authors have noted a striking correlation between the presence of salt lakes at the surface and the occurrence of subsurface preglacial and englacial valleys in the region (e.g., Witkind, 1952; Rueffel, 1968; Freeze, 1969), and have surmised that these buried valleys act as conduits for water and ions to the salt lakes. Although details of the groundwater hydrodynamics and the interaction of groundwater with individual salt lake basins in the northern Great Plains are poorly studied and understood, variation in subsurface water composition is reasonably well documented on a regional basis (e.g., Brown, 1967; Lennox et al., 1988). Most of the groundwater in unconsolidated surficial deposits is of low to moderate salinity (<3 ppt TDS) and dominated by calcium, magnesium, and bicarbonate ions. In western Saskatchewan and eastern Alberta, this shallow-drift groundwater is usually characterized by sulfate rather than bicarbonate ions. The shallow bedrock aquifers (Upper Cretaceous and younger rocks) are mainly sodium-bicarbonate in southern Alberta, calcium-magnesium-sodium-sulfate in Saskatchewan, and calcium-magnesium-sodium-bicarbonate in western Manitoba. The deeper Paleozoic and Cenozoic bedrock contains much higher salinity water (up to 300 ppt TDS) that is usually dominated by sodium and chloride.

The lakes

> In this region, there are numerous ponds and small lakes in the hollows among the hills, most of them being more or less brackish or nauseous to the taste from the presence of sulphates of magnesia and soda and other salts. During the dry season of autumn, the water evaporates completely from many of these ponds leaving their beds covered by the dry white salts, which look like snow and are blown about in the wind. (Bell, 1875)

In most of the northern Great Plains and Palliser's Triangle, ponded saline and hypersaline brines are the only surface waters present. As a group, the lakes of this region are unique: there is no other area in the world that can match the concentration and diversity of saline lake environments exhibited in the western interior region of Canada and northern United States. The immense number of individual salt lakes and saline wetlands in this region of North America is staggering. Estimates vary from about 1.5 million to greater than 10 million, with densities in some areas being as high as 120 lakes/km^2 (Last, 1989a). As shown in Figure 2, the vast majority of these lakes in Palliser's Triangle are small, shallow, and ephemeral (i.e., playas).

Despite nearly a century of scientific investigation of these salt lakes, we have, in the last two decades only, advanced far enough to appreciate the wide spectrum of basin types, water chemistries, and geolimnological processes that are operating in the modern settings. Hydrochemical data are available for about 500 lake brines in the region. Mineralogical, textural, and geochemical information on the modern bottom sediments has been collected for just over 100 of these lakes. The stratigraphic records of only twenty of the basins in the entire northern Great Plains of both Canada and United States have been examined, with just eight of these from Palliser's Triangle per se. Complete sequences of the entire Holocene have been reported from only three of these lakes.

The lake waters show a considerable range in ionic composition and concentration. Early investigators, concentrating on the most saline brines, emphasized a strong predominance of Na^+ and SO_4^{2-} in the lakes (Cole, 1926; Sahinen, 1948; Govett, 1958). It is now realized, however, that not only is there a complete spectrum of salinities from less than 1 ppt TDS to over 400 ppt, but also virtually every water chemistry type is represented in lakes of the region (Fig. 3). Rawson and Moore (1944), Rutherford (1970), Hammer (1978), and Lambert (1989) have compiled lake-water chemistries in the region. Details of spatial trends and regional variations in the lake-water composition in adjacent areas of northern United States have been discussed by Gorham et al. (1983) and Winter (1977), and in western Canada by Last and Schweyen (1983), and Last (1989a, 1988). Lake brines with the highest propor-

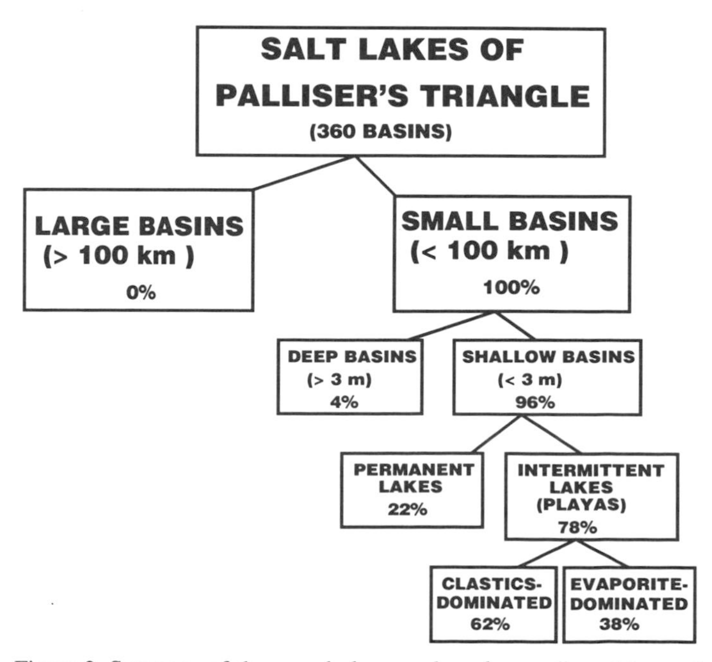

Figure 2. Summary of the morphology and modern sediment type of 360 surveyed lakes in Palliser's Triangle.

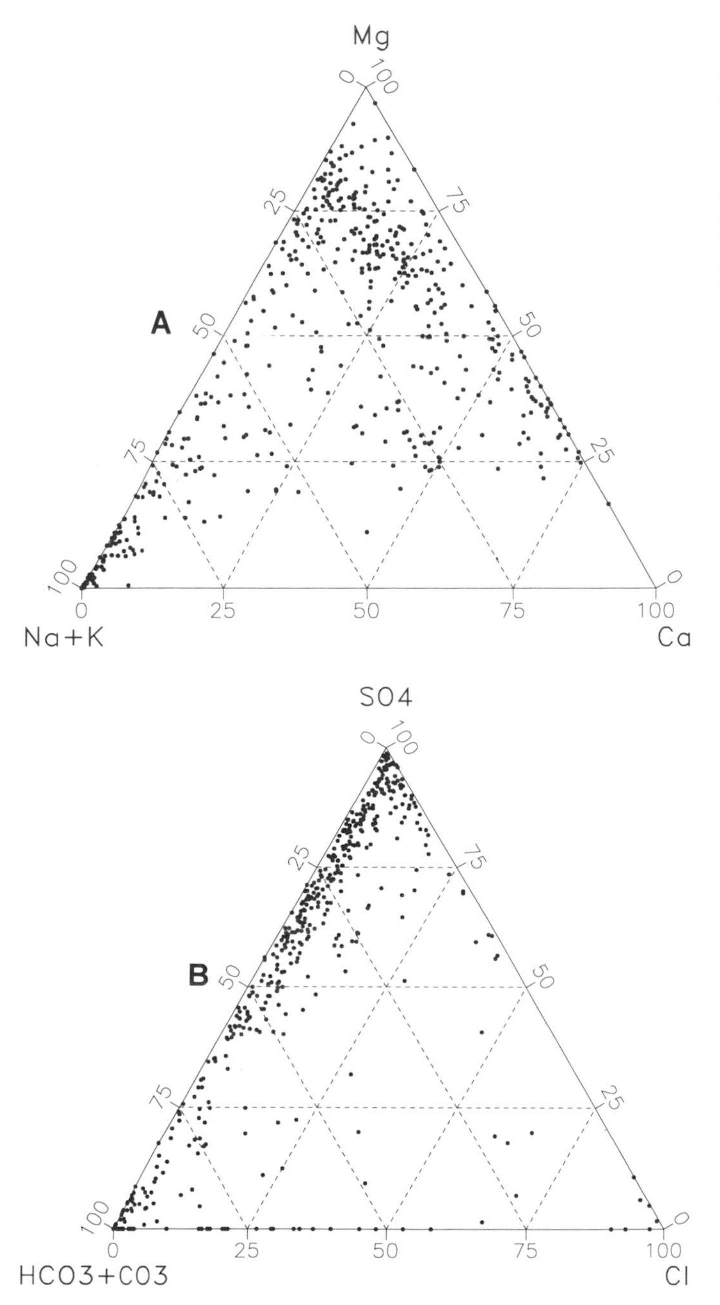

Figure 3. Trilinear diagrams showing (A) cation and (B) anion ratios in 360 surveyed lakes in Palliser's Triangle.

tions of sodium and sulfate ions generally occur in east-central Alberta and west-central Saskatchewan, whereas calcium and carbonate-rich brines dominate in the north and east part of the region. Brines with relatively high chloride and magnesium contents occur in western and central Manitoba.

With such a vast array of compositions, it is difficult to generalize. Nonetheless, the paucity of Cl-rich lakes makes the basins of Palliser's Triangle somewhat unusual compared with salt lakes in many other areas of the world (e.g., Australia, western United States). Significant short-term temporal variations in the brine composition, which can have important effects on the composition of the modern sediments, have been well documented in several individual playa basins (Hammer, 1990, 1986, 1978; Last, 1989b, 1984; Lieffers and Shay, 1983; Rozkowska and Roskowski, 1969). For example, the water in Ceylon Lake, a salt playa located about 100 km south of Regina, is dominated by sodium, sulfate, and bicarbonate ions during early spring but becomes a magnesium-chloride-sulfate brine by late summer.

From a sedimentological/mineralogical perspective, the wide range of water chemistries exhibited by the lakes results in an unusually large diversity of modern sediment compositions. Over 40 species of endogenic precipitates (i.e., originating directly from the lake water) and authigenic minerals (i.e., derived by diagenetic alteration of previously deposited sediment, or by direct precipitation from pore-water solutions) have been identified in the lacustrine sediments (Last, 1989a; Last and Slezak, 1987). The most common nondetrital components of the modern sediments include: calcium and calcium-magnesium carbonates (magnesian calcite, aragonite, dolomite), and sodium, magnesium, and sodium-magnesium sulfates (mirabilite, thenardite, bloedite, epsomite). Many of the basins whose brines have very high Mg/Ca ratios also have hydromagnesite, magnesite, and nesquehonite. Unlike salt lakes in many other areas of the world, halite, gypsum, and calcite are relatively rare endogenic precipitates in lake sediments of the Great Plains.

Sediment accumulation in these salt lakes is controlled and modified by a wide variety of physical, chemical, and biological processes. Smoot and Lowenstein (1991), Warren (1989), Allen and Collinson (1986), Kendall (1984), Eugster and Kelts (1983), Eugster and Hardie (1978), and Hardie et al. (1978) provide discussions and comprehensive overviews of the general suite of processes operating in salt lakes. The details of the many modern sedimentary processes can be exceedingly complex and difficult to discuss in isolation. In broad terms the processes operating in the salt lakes of the Great Plains are ultimately controlled by three basic factors or conditions of the basin: (a) basin morphology, (b) basin hydrology, and (c) water salinity and composition. Combinations of these parameters interact to control nearly all aspects of modern sedimentation in these lakes and give rise to four "end member" types of modern saline lacustrine settings in the region: (1) shallow lakes (playas) dominated by clastic sediment, (2) shallow lakes (playas) dominated by chemically precipitated sediment, (3) deep water (perennial) lakes dominated by clastic sediment, and (4) deep water (perennial) lakes dominated by chemically precipitated sediment (Fig. 4).

Table 1 summarizes the dominant nonbiological processes operating in the four types of basins. Two fundamental processes common to all of the lakes in the region are (1) the acquisition of water, sediment, and solutes by direct precipitation, river and stream inflow, and groundwater influx; and (2) the concentration of solutes by evaporation. The playa basins are further affected by a distinct suite of physical and chemical

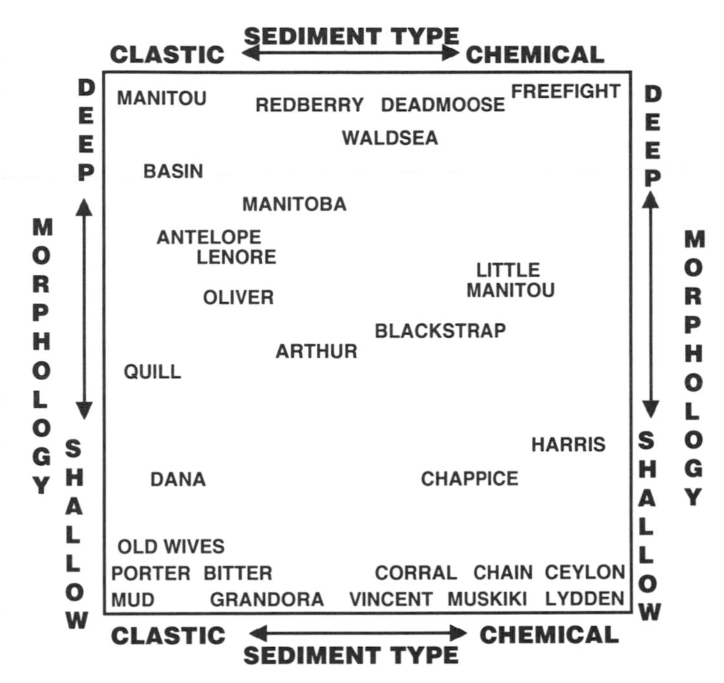

Figure 4. Schematic diagram showing the range of lacustrine settings in the northern Great Plains on the basis of basin morphology and modern sediment type. See Figure 1 for locations of representative lakes identified.

processes dominated by seasonal (or periodic) flooding and desiccation, influx of clastic debris by sheetflood and wind, precipitation of soluble and sparingly soluble salts, formation of detrital vegetation mats and microbialites, deflation, and sediment disruption by various cryogenic, biogenic, and chemical processes. In contrast, the deep-water basins are characterized by a suite of processes dominated by seasonal (or rhythmic) carbonate mineral precipitation and detrital sedimentation, formation of subaqueous salt cumulates by evaporative concentration and freeze-out precipitation, sulfate reduction and sulfide mineral precipitation, shoreline deposition and erosion, turbidity flow, and pelagic fallout.

Although overlap does occur, the sedimentological controls, the biological influences, and the diagenetic susceptibility of these four basic lake types differ dramatically. Therefore, the resulting sedimentary facies are (or should be) readily distinguished from one another in a stratigraphic succession as shown in Figure 5. How well this distinction can be made in the paleorecord of a given basin is a direct reflection of our level of understanding about each modern end-member type. Even though today the number of salt-dominated playas is small relative to the clastics-dominated basins, the former have received much more attention because they constitute the basis of a $\$40 \times 10^6$ yr^{-1} salt minerals industry in the Great Plains (Broughton, 1984). Thus, our knowledge about the processes and our ability to recognize and interpret the stratigraphic signals within the salt-dominated playa basins is high relative to other saline lake types in the region.

TABLE 1. DOMINANT PHYSICAL AND CHEMICAL PROCESSES AFFECTING SEDIMENTATION IN LAKES OF THE NORTHERN GREAT PLAINS*

Processes important in all the lakes
Influx of water, sediment, solutes from streams, groundwater, direct precipitation, sheetwash.
Evaporative concentration of water.

Processes important in deep-water, clastics-dominated lakes
Development of thermal stratification of water column.
Turbidity flow, interflow.
Flocculation of fine-grained material.
Cyclic and rhythmic sedimentation.
Shoreline erosion, deposition.
Delta sedimentation.
Aeolian sediment influx.

Processes important in deep-water, salt-dominated lakes
Development of meromixis.
Bio-mediated carbonate precipitation.
Evaporative carbonate precipitation.
Formation of subaqueous salt cumulates.
Solute concentration by formation of ice cover.
Freeze-out precipitation of salts.
Sulfate reductation, sulfide mineral precipitation.
Cyclic and rhythmic sedimentation.
Clay mineral authigenesis.

Processes important in shallow-water, clastics-dominated lakes
Cyclic flooding, desiccation.
Deposition, erosion by sheetflood flow.
Aeolian influx, deflation.
Intrasedimentary salt precipitation.
Formation of efflorescent crusts.
Formation of vegetation mats.
Formation of subsurface salt cements.
Pedogenesis.
Sediment disruption by freeze-thaw, bioturbation.

Processes important in shallow-water, salt-dominated lakes
Cyclic flooding, desiccation.
Wind set-up, wind-controlled localization of salt precipitation.
Precipitation of salts at air-water interface.
Formation of crystal rafts and aggregates.
Evaporative pumping. Formation of efflorescent crusts.
Formation of carbonate hardgrounds, crusts.
Formation of microbialites.
Development of meromixis.
Subaqueous cumulate and bottom salt precipitation.
Formation of salt spring deposits.
Formation of rounded accretionary salt grains.
Reworking and distribution of clastic salts.
Temperature-induced mineral transformations, phase changes.
Formation of salt cements.
Salt karsting.
Mud diapirism, reworking of fine-grained clastic material.

*Modified from Last and Schweyen, 1983.

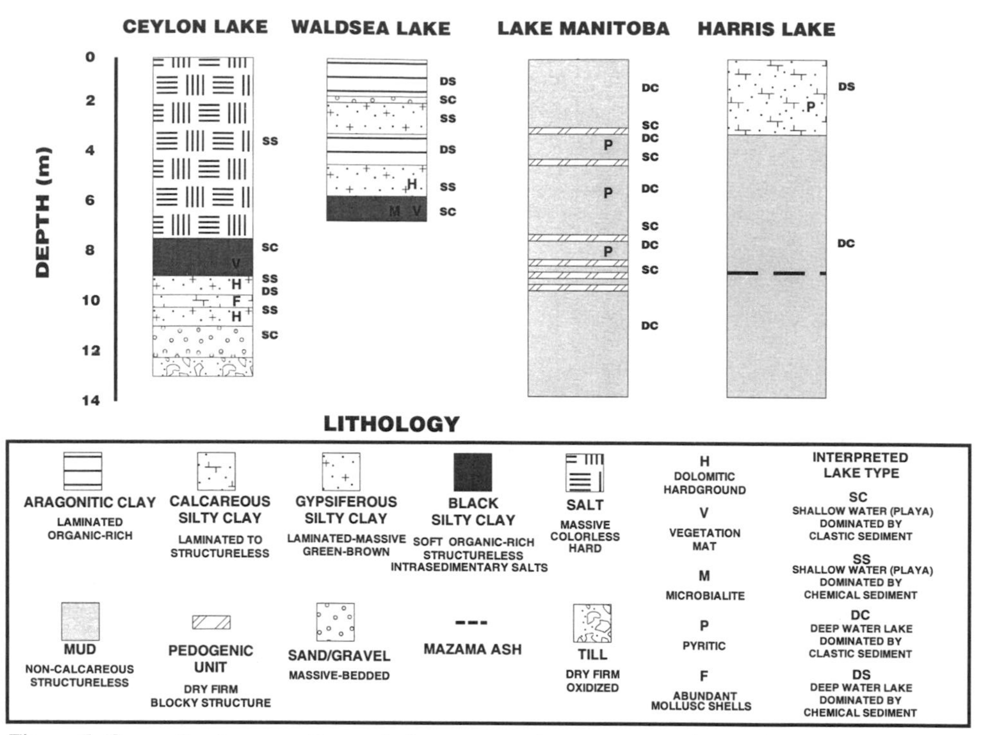

Figure 5. Generalized, composite vertical sequences from representative well-studied lakes in the northern Great Plains. Ceylon Lake from Last (1990); Waldsea Lake from Last and Schweyen (1985); Lake Manitoba from Teller and Last (1981); Harris Lake from Last and Sauchyn (1993).

Finally, despite the fact that playas and shallow salt lakes have been the most extensively studied saline terrestrial environments, not only in the Plains region of North America but also on a worldwide basis (Eugster and Kelts, 1983), it is commonly held that these types of basins provide a less-than-desirable record of paleoenvironmental conditions (e.g., Street-Perrott and Harrison, 1985; Hammer, 1986). Because playas in arid and semiarid regions undergo periodic and frequent subaerial exposure, the likelihood of significant post-depositional alteration of the sediment is great and the loss of record through erosion, nondeposition, and deflation is to be expected. Clearly the level of resolution that can be achieved in some deep-water, permanently stratified basins cannot usually be matched in the ephemeral playa environment. However, such is the nature of paleohydrological investigations in the Great Plains that: (a) most of the lakes are, indeed, shallow playa basins; and more importantly, (b) nearly all of the lakes in the region, both deep and shallow, for which we do have stratigraphic information have experienced shallow-water, playa phases at some point in their Quaternary history (Schweger and Hickman, 1989; Last and Slezak, 1988).

PROBLEMS AND PITFALLS

Mining industry activity

The playas and saline wetlands of the northern Great Plains serve a great variety of uses. Many studies have documented the importance of these saline terrestrial environments to surface runoff and flow stabilization, erosion control, waste assimilation, agriculture, and wildlife habitat (e.g., Waite, 1980, 1986; Adams, 1988; Richardson and Arndt, 1989). The economic significance of some playa lakes is further enhanced by the presence of large quantities of economically valuable industrial minerals. A form of sodium sulfate known as Glauber's salt has been mined and commercially extracted from these lakes for over 70 years. In the last two decades the dollar value of this industrial mineral produced from the lakes in western Canada has increased fivefold. Associated with this increase in market value is an increase in tonnage produced and, to meet the demand, an increase in the number of lakes being mines. Historically, production has occurred from 28 separate basins in the region, with 9 playas being mined today. There are more than 60 basins in Palliser's Triangle with proved commercial reserves of Na_2SO_4 greater than 100,000 tonnes (Slezak and Last, 1985). Magnesium salts are also a potential commercial product from a number of the saline lakes (Tomkins, 1954).

Although mining of these industrial minerals from the lakes occurs in various ways (e.g., solution mining, excavation, precipitation from brine), the end result of the extraction process(es) on the basin is, from the standpoint of paleoenvironmental research, undesirable. Because the lakes that are the most economically attractive for mineral exploitation are often also those whose sediments provide some of the best evidence for geochemical and brine compositional changes and water level fluctuations through time, the mining of nearly 40 of the salt lakes, and the likelihood of expanded mining efforts in the near future, becomes a major concern for paleohydrological work in Palliser's Triangle.

Metiskow Lake example. An example of irretrievable loss of valuable paleoenvironmental data due to mining activities is that of Metiskow Lake, Alberta. The economic potential of this playa lake in eastern Alberta was first recognized by Cole (1926) who calculated a sodium sulfate reserve in the lake of about 5×10^6 tonnes. Metiskow is unusual in that its brine is dominated by bicarbonate ions, rather than sulfate. Subsequent drilling and bulk chemical analyses of the sediments in the lake by various companies during the 1960s confirmed that the 17-m-thick lacustrine salt sequence contained some 250,000 tonnes of sodium carbonate in addition to the large sodium sulfate reserve.

Before mining operations obliterated the stratigraphic record in the basin, Wallick and Krouse (1977) and Wallick (1981) were able to decipher the groundwater flow and likely water-rock chemical reactions that gave rise to this unusual deposit. These two studies, as important as they are in providing quantitative information on the source of the salts, merely serve to accentuate the tremendous potential of the paleoenvironmental record in Metiskow. The basin contained a thick sequence of alternating sodium carbonate salts, sodium and calcium sulfates, and detrital material; similar mineral assemblages have been used very effectively in other salt-lake sequences to reconstruct the details of paleohydrochemical conditions and changes (e.g., Eugster and Smith, 1965). The magnitude of this lost opportunity is further emphasized by the report of a single ^{14}C date of 10,250 yr B.P. (Wallick, 1981) from the base of the salt section, which indicates that Metiskow was one of the few lakes cored in the region to date that may have provided a complete record of Holocene and late Pleistocene environmental change.

Postdepositional changes

Numerous summary and overview papers have already stressed the susceptibility of salt-lake sediments, particularly evaporites, to post-depositional changes. As a generalization, the playa basins are more prone to post-depositional changes than perennial lakes because of their periodic desiccation and subaerial exposure. In the playa lakes of the northern Great Plains, three major types of diagenetic alteration have been documented: (1) saline mineral diagenesis, (2) salt dissolution and karsting, and (3) diapirism.

Mineral diagenesis. Evaporites of salt lakes of the Great Plains have a distinct and characteristic mineralogy. It is this mineralogy that can be so useful in helping to decipher paleochemistry of the brines and paleohydrology of the basins (Teller and Last, 1990). Regardless of precisely *how* a certain evaporite formed (e.g., by "simple" concentration of lake water via isothermal evaporation, by a temperature increase or decrease of the brine, by mixing of brines of different compositions, by biologically induced precipitation, etc.), the presence of that particular salt mineral in a stratigraphic sequence implies the formative brine was saturated (or supersaturated) with respect to that precipitate. Thus, a given suite of evaporite minerals can be used to calculate, in considerable detail, the thermodynamic conditions of the evaporating solution. Problems arise, however, if it cannot be assumed that the minerals are either endogenic or at least very early diagenetic (formed essentially at the sediment-water interface). In many salt lakes, dense, hypersaline surface brines percolate downward into porous and permeable subsurface sediments. Because these dense brines are late-stage residual products, they often have dramatically different chemical compositions than the normal lake or groundwater. The ability of these descending brines to radically alter the original mineral composition of large sections of the subsurface sediments has been well documented (Sonnenfeld, 1984; Warren, 1989).

Salt dissolution and karsting. In addition to conversion of one evaporite mineral to another, undersaturated groundwater or surface water can completely dissolve the most soluble components of the stratigraphic record of a playa lake. The formation of deep karst chimneys and large dissolution pits is occurring today in numerous saline playa basins in Palliser's Triangle. For example, in Ceylon Lake, a salt-dominated playa located in south-central Saskatchewan, saline karst chimneys up to 9 m deep have been identified (Last, 1989b). In Lydden Lake, another salt playa located west of Saskatoon, Saskatchewan, solution pits up to 30 m wide and 3 m deep are present. In still other basins, the occurrence of buried mud "mounds", highly irregular salt thicknesses (Fig. 6), and abrupt compositional changes over short distances (Fig. 7) indicate that the process of salt removal has also taken place in the past. It is especially important to recognize this when attempting to interpret the stratigraphic record of the playas because large vertical sections of the lacustrine sediment can be affected.

Mud diapirism. Diapirism, or the process of piercing an overlying geological unit by an underlying mobile core material, is a phenomenon commonly described in several modern and recent sedimentary environments. Mud diapirs, or mudlumps, are particularly abundant in birdfoot-type deltas, where, for example in the Mississippi River Delta, they can penetrate 100 m of delta-front sands (Coleman and Prior, 1980). Features of similar origin also occur in saline playas of the northern Great Plains (Last, 1984). Although the precise mechanisms of the diapirism are not yet known, these features in the salt lakes are likely the result of an instability brought about by the loading of relatively low density, water saturated lacustrine muds by a thick section of dense and nonpermeable salt. This instability causes the muds to flow upward through the salt to the lake surface. At the surface, this clastic sediment is redistributed by wave action and incorporated into the modern deposits of the lake thereby leading to potential contamination.

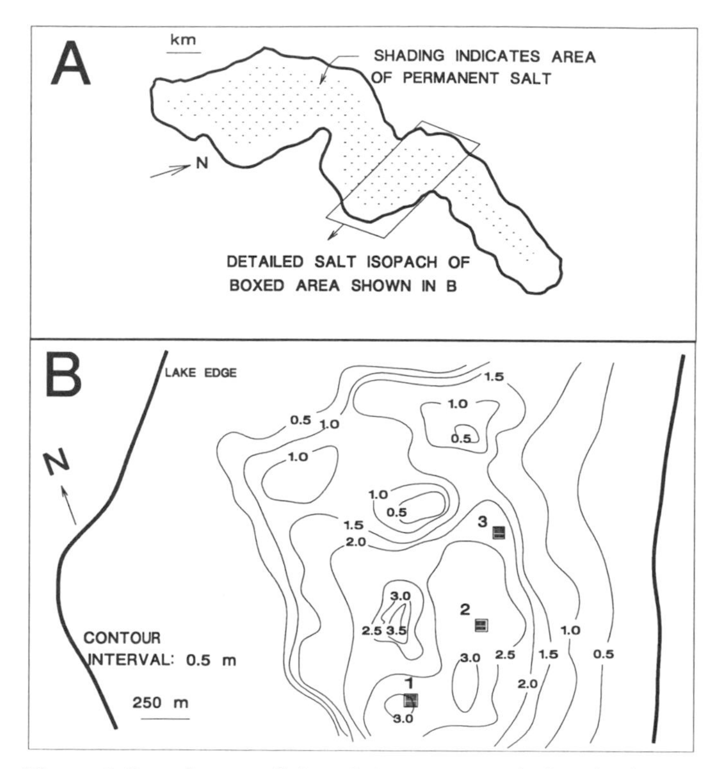

Figure 6. Isopach map of the salt in an unnamed playa basin near Verlo, Saskatchewan, in Palliser's Triangle showing dramatic fluctuations in thickness due to active and paleo-salt karst development. The isopach map in B was constructed on the basis of a grid of auger hole data at approximately 250 m spacing. The rectangles numbered 1, 2, and 3 in B are the locations of cores shown in Figure 7.

Incomplete understanding of sedimentary and geochemical processes

As discussed in the previous section, a complete understanding of the sedimentary processes operating in the entire spectrum of salt lakes of the Great Plains is still far off. Because virtually all geological interpretations of the stratigraphic records in these basins depend on a knowledge of the modern environments and modern sedimentary processes, it is important that these deductions be made with great care. Following are several of the more important, but still unresolved "conceptual" problems that continue to obscure the paleohydrological interpretation of the sediment record in these playas.

Significance of fine-grained clastic units. In a sedimentary basin strongly influenced by physical factors such as wind-generated waves, currents, and river inflow, the distribution and character of the accumulating sedimentary material is controlled to a major degree by the level of energy at the depositional site. Thus, coarse-grained sediments (sands, gravels) are generally interpreted to have been deposited under relatively shallow water, high energy conditions (e.g., a beach or delta), whereas very fine grained sediment (clay-sized material) requires settling through a deep, essentially motionless water mass. As stressed by Sly (1978), Håkanson and Jansson (1983), Rust and Nanson (1989), and Teller and Last (1990), many other factors can interfere with this simple grain size–water depth–energy relationship, including flocculation of fine-grained material, formation of pedogenic aggregates within the watershed, basin morphology, weathering and erosion characteristics of the watershed, and vegetation cover.

Thin beds and laminae of fine-grained siliciclastic material commonly occur interbedded with evaporites in the stratigraphic records of the salt-dominated playas of the Great Plains. Their paleohydrological significance and sedimentological interpretation are particularly problematic. They may represent periods of relatively deep and fresh water conditions in which increased runoff from the surrounding watershed (the result of a more humid climate) brought an increased influx of clastic material to the lake. Conversely, the fine silt and clay units could be residual products, the end result of dissolution and downward leaching of the salts due to prolonged desiccation of the playa and a lowered groundwater table induced by a more arid climate. The fine clastics may also have been deposited during periods of increased aeolian activity associated with increased aridity, or carried into the basin as sand-sized pedogenic aggregates by sheet floods. Obviously, each of these possible interpretations represents significantly different hydrological conditions in the basin.

Significance of carbonate mineralogy. Carbonate minerals are among the most common and also the most useful inorganic constituents of lakes sediments. Many studies have documented the suitability of calcite and aragonite in paleolimnology (e.g., Kelts and Hsü, 1978; Dean, 1981; Eugster and Kelts, 1983; Dean and Fouch, 1983; Behbehani et al., 1986). The application of other carbonate mineral species, such as dolomite, magnesite, and siderite, which commonly occur in salt lake sediments, has been less widespread but does offer considerable potential (e.g., Rosen et al., 1988).

Based mainly on observational data provided by Müller et al. (1972), conventional wisdom holds that the mineral sequence of: low-Mg calcite → high-Mg calcite → aragonite → dolomite → magnesite/huntite reflects increasing Mg/Ca ionic ratios in the precipitating solution (Fig. 8) and, most likely, increasing salinities. Thus, it has become commonplace to use the stratigraphic variation of these species in a lacustrine basin to deduce past ionic ratios and salinity of the lake water (e.g., Müller and Wagner, 1978; Last, 1982; Last and Schweyen, 1985; Allen and Collinson, 1986; Vance et al., 1993). Unfortunately, there are many possible sources of error associated with this cookbook approach. In order to be used the carbonates must be of primary origin and must be in situ precipitates. However, it is commonly difficult to distinguish primary endogenic carbonates from either primary or secondary diagenetic precipitates. Talbot and Kelts (1986, 1990) and Talbot (1990) discuss the complexity of the interpretations if any of the carbonate components are of diagenetic origin. Furthermore, other factors besides, or in addition to, Mg/Ca and salinity have a

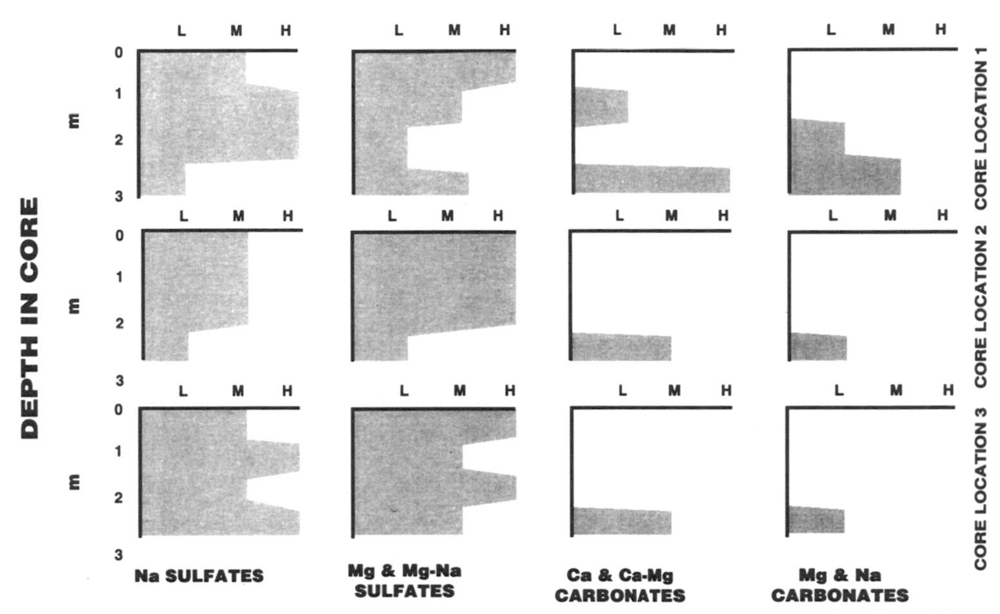

Figure 7. Summary of fluctuations in relative abundance of the major evaporite and endogenic/authigenic carbonate mineralogy in cores from the un-named lake shown in Figure 6. L, M, and H stand for low, medium, and high relative abundance. Diagram is generalized from semiquantitative X-ray diffraction analysis data of samples taken at appproximately 20 cm spacing in the cores. The core locations are about 500 m apart.

major influence on the specific carbonate species being precipitated. For example, both the alkalinity and the sulfate concentration of the water are thought to be key factors controlling the formation of dolomite (Morrow, 1982; Baker and Kastner, 1981; Kastner, 1986). Temperature of the solution, in addition to the Mg/Ca ratio, plays a major role in dictating how much Mg^{2+} is incorporated into the calcite lattice in the formation of high-Mg calcite (Müller and Wagner, 1978).

Importance of saline mineral metastability. The two fundamental assumptions that allow sedimentologists to deduce the composition of a brine from the mineral record preserved in the lake sediment are that: (a) the composition of the mineral suite has not been significantly altered by post-depositional changes, and (b) the mineral suite that is ultimately preserved is a true reflection of the water composition that existed in the basin at the time of deposition. Problems associated with (a) have already been mentioned. If the saline playa is viewed simply as a container of water in which relatively dilute inflow is concentrated by evaporation to the point of saturation with respect to the minerals present, assumption (b) is valid: the mineralogy is, indeed, a reasonable reflection of the brine composition at the point of precipitation. Unfortunately, this view of a playa or any salt lake as a container of water undergoing evaporation is an oversimplification of the real world. Apart from ignoring the role of biological processes in mineral formation, this approach is inadequate for at least two major geochemical reasons: Firstly, the newly formed minerals are not necessarily isolated from the brine immediately upon precipitation. This ability of the salts to react with a changing brine composition on a seasonal basis (as is the case in many playa basins) or as the precipitate settles through a thermally or chemically stratified water column greatly increases the likelihood of mineral alteration even before being incorporated into the sediment record. Secondly, this simple view does not take into account metastability of the salts. The

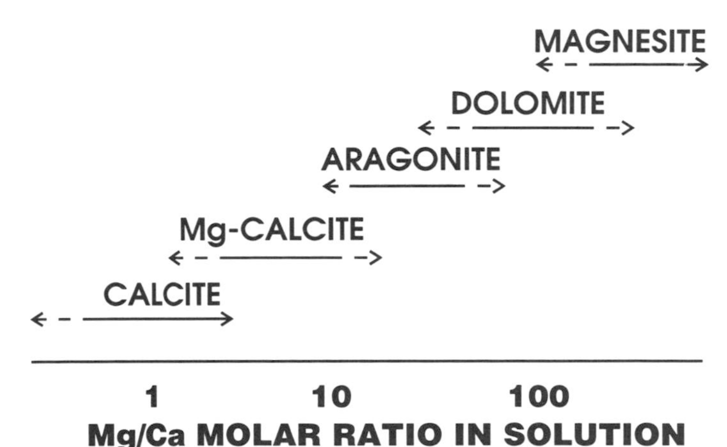

Figure 8. Summary of commonly cited ranges of Mg/Ca ratios in carbonate-precipitating solutions relative to the stable Ca-Mg carbonate mineral.

formation of metastable phases is a common phenomenon in low-temperature geochemistry, particularly in late-stage evaporation of brines (Krauskopf, 1979; Holser, 1979; Drever, 1988). Although with time, the metastable phases should transform to thermodynamically stable phases, this transformation can be slow enough to allow the metastable material to persist in the geological record, thereby muddling the interpretation of the mineral suite.

FUTURE PALEOHYDROLOGICAL RESEARCII DIRECTIONS

Where does playa lake paleohydrology in the northern Great Plains go from here? The multidisciplinary nature of paleoenvironmental work being undertaken on lakes today makes it difficult to predict future development of the science. However, within the geographic and topical constraints of this paper, the axiom "more, better, faster" does apply. I do not think that we should be too cynical about attempting to "do" paleohydrology without first knowing everything about the modern lakes. Sedimentologists have long accepted the fact that much of what is known about some depositional settings is based not on the modern environments but rather on the preserved ancient record. A high priority must, however, be placed on overcoming the most crucial deficiencies in present sedimentological and geochemical concepts/knowledge as summarized in the previous section: quite simply, *more* modern sedimentological data must be collected from the lakes, especially from the clastics-dominated playas. At the same time, however, many *more* salt lake basins have to be cored, with particular emphasis in areas where there is little data today (i.e., most of western and southwestern Saskatchewan and eastern/southern Alberta) and in strategically significant areas, such as near present-day climatic/vegetational boundaries.

The quality of sedimentological/geochemical data has to be improved. This is particularly important in the collection of mineralogical data; for example, it is much more useful from a paleochemical perspective to know that the stratigraphic unit contains bloedite than merely that it contains traces of magnesium. Our understanding of the evolution of salt lakes in the western interior of Canada may be quite different today had the large amount of data collected in conjunction with industrial development of the Metiskow Lake sodium carbonate-sulfate deposit been placed in a mineralogical framework, rather than a bulk chemistry context.

Finally, there has been no better time than now to begin this new wave of paleohydrological/paleoenvironmental investigation. Regional climatic changes, drought, salinization, and sustainable development have all become "buzz words" of the 1990s. Local, provincial, and federal governments as well as international scientific bodies have rapidly become aware of the need for documentation of past environmental changes in order to properly evaluate present-day trends and fluctuations.

In many ways the study of lake sediments in most of western Canada has remained relatively simple and traditional, more or less unchanged from the 1960s and 70s. The next decade will likely bring a great increase in sophistication of methodology as well as technology. Although the general evolutionary histories of several playa basins in the region are reasonably well known, because of recent technological advances, the potential for deducing much more explicit paleohydrological information from the sediment records is still high. Today it is possible to analyze, rapidly and with high confidence, the crystal size distribution and crystal-shape characteristics of the endogenic and authigenic material in a single submillimeter lamina of sediment core. It is also now possible to determine the bulk chemical composition, the trace-element content, the stable-isotopic composition, and even the zoning characteristics of single micron-sized crystals from one of these laminae. Such detailed investigations, unthinkable just a decade ago, would almost certainly significantly advance the understanding of playa sedimentology, evolution of mineral-brine systems through time, and paleoclimatic/hydrologic fluctuations in this region. The recent advances in geochronology and sample acquisition discussed by Smol (1989) have yet to be applied to playa lakes of the region; these also will help to revolutionize the paleohydrological efforts in Palliser's Triangle.

SUMMARY AND CONCLUSIONS

Palliser's Triangle encompasses a large portion of the northern Great Plains of Canada. It is a 200,000-km^2, semiarid to arid region in which geology, climate, and hydrology have interacted to form many saline lakes. Although there is a wide spectrum of basin morphologies, sedimentary characteristics, and water compositions and concentrations, the modern lakes can generally be pigeon-holed into one of several basic types: (1) clastics-dominated playa; (2) salt-dominated playa; (3) deep-water, perennial lake dominated by clastic sediment; and (4) deep-water, perennial lake dominated by chemically precipitated sediment. The playa basins vastly outnumber the perennial lakes in the region. The level of understanding of the physical, chemical, and biological processes operating in these lakes, and the resulting sedimentary facies in the basins, is incomplete and uneven. Salt-dominated playas have received most of the past sedimentological and paleohydrological study.

Three major problems and pitfalls can be identified in using the stratigraphic records of playas to help interpret the paleohydrology of the basins and hydrological changes that occurred in Palliser's Triangle. (1) The playas of the region support an active and growing salt minerals industry. Unfortunately, often the lakes that are most attractive for mining are also the ones that contain the best potential for paleohydrological research. (2) A variety of post-depositional changes can significantly alter the composition and nature of the preserved stratigraphic records in these playa lakes, thereby obscuring the paleorecord. (3) Our incomplete understanding of several basic sedimentary processes and genesis of some commonly occurring stratigraphic units in these basins is limiting the precision of our interpretations.

There are no conclusions to this paper, only several words of encouragement and caution. The scientific study of playa lakes in the northern Great Plains is ripe for an explosive expansion. The paleoenvironmental problems are difficult but not intractable. The tools are available. The potential payoffs are big. However, we must guard against resorting to excessive use of comparison and analogue study. The sedimentary realm of the playa environments of the Great Plains is, in many ways, unique. What is fashionable in Australia or what "works" in southwestern United States, for example, may not necessarily be applicable in the Great Plains.

ACKNOWLEDGMENTS

This paper was inspired by the excellent and thought-provoking overview contributions of Frey (1988) and Smol (1989) and by many discussions of paleohydrology with my colleagues and students at University of Manitoba. I thank J. P. Smol, A. C. Kendall, and an anonymous reviewer for their critical comments on an earlier draft. Alberta Sulphate Limited is thanked for giving me permission to examine their files and unpublished data on the Metiskow deposit.

REFERENCES CITED

Adams, G. D., 1988, Wetlands of the Prairies of Canada, *in* National Wetlands Working Group, Canada Committee on Ecological Land Classification. Wetland of Canada: Ecological Land Classification Series, no. 24, p. 155–198.

Allen, P. A., and Collinson, J. D., 1986, Lakes, *in* Reading, H. G., ed., Sedimentary environments and facies (second edition): Oxford, Blackwell Scientific Publications, p. 63–94.

Baker, P. A., and Kastner, M., 1981, Constraints on the formation of sedimentary dolomite: Science, v. 213, p. 214–216.

Behbehani, A., and 6 others, 1986, Sediments and sedimentary history of Lake Attersee (Salzkammergut, Austria): Hydrobiologia, v. 143, p. 233–246.

Bell, J. S., 1875, Report on the country between Red River and the South Saskatchewan: Geological Survey of Canada Report of Progress 1873–74, p. 66–93.

Broughton, P. L., 1984, Sodium sulphate deposits of western Canada, in Guillet, G. R., and Martin, W., eds., The geology of industrial minerals in Canada: Canadian Institute of Mining and Metallurgy Special Volume 29, p. 195–20.

Brown, I. C., ed., 1967, Groundwater in Canada: Geological Survey of Canada Economic Geology Report 24, 228 p.

Cole, L. H., 1926, Sodium sulfate of western Canada. Occurrence, uses and technology: Canadian Department of Mines Publication 646, 155 p.

Coleman, J. M., and Prior, D. B., 1980, Deltaic sand bodies: American Association of Petroleum Geologists Continuing Education Course Note Series, no. 15, 171 p.

CNC/IHD (Canadian National Committee for the International Hydrologic Decade), 1978, Hydrologic atlas of Canada: Ottawa, Fisheries and Environment Canada, 75 p.

Dean, W. E., 1981, Carbonate minerals and organic matter in sediments of modern north temperate hard-water lakes, *in* Ethridge, F. G., and Flores, R. M., eds., Recent and ancient nonmarine depositional environments: Models for exploration: Society of Economic Paleontologists and Mineralogists Special Publication 31, p. 213–231.

Dean, W. E., and Fouch, T. D., 1983, Chapter 2: Lacustrine, *in* Scholle, P. A., Bebout, D. G., and Moore, C. H., eds., Carbonate depositional environments: American Association Petroleum Geologists Memoir 33, p. 98–130.

Drever, J. I., 1988, The geochemistry of natural waters (second edition): Englewood Cliffs, New Jersey, Prentice Hall, 437 p.

Eugster, H. P., and Hardie, L. A., 1978, Saline lakes, *in* Lerman, A., ed., Lakes: Chemistry, geology, physics: New York, Springer-Verlag, p. 237–293.

Eugster, H. P., and Kelts, K., 1983, Lacustrine chemical sediments, *in* Goudie, A. S., and Pye, K., eds., Chemical sediments and geomorphology: London, Academic Press, p. 321–368.

Eugster, H. P., and Smith, G. I., 1965, Mineral equilibria in the Searles Lake evaporites, California: Journal of Petrology, v. 6, p. 473–522.

Freeze, R. A., 1969, Regional groundwater flow—Old Wives Lake drainage basin, Saskatchewan: Department of Energy Mines and Resources Inland Waters Branch Scientific Series 5, 245 p.

Frey, D. G., 1988, What is paleolimnology?: Journal of Paleolimnology, v. 1, p. 5–8.

Gorham, E., Dean, W. E., and Sanger, J. E., 1983, The chemical composition of lakes in the north-central United States: Limnology and Oceanography, v. 28, p. 287–301.

Govett, G.J.S., 1958, Sodium sulfate deposits in Alberta: Alberta Research Council Preliminary Report 58-5, 34 p.

Håkanson, L., and Jansson, M., 1983, Principles of lake sedimentology: New York, Springer-Verlag, 316 p.

Hammer, U. T., 1990, The effects of climate change on the salinity, water levels and biota of Canadian prairie saline lakes: Verhandlung des Internationalen Vereins der Limnologie, v. 24, p. 321–326.

Hammer, U. T., 1986, Saline lake ecosystems of the World: Dordrecht, The Netherlands, Dr. W. Junk Publications, 616 p.

Hammer, U. T., 1978, The saline lakes of Saskatchewan, Ill. Chemical characterization: International Revue der Gesamten Hydrobiologie, v. 63, p. 311–335.

Hardie, L. A., Smoot, J. P., and Eugster, H. P., 1978, Saline lakes and their deposits: A sedimentological approach, *in* Matter, A., and Tucker, M. E., eds., Modern and ancient lake sediments: International Association of Sedimentologists Special Publication Number 2, p. 7–42.

Holser, W. T., 1979, Mineralogy of evaporites, *in* Burns, R. G., ed., Marine minerals: Mineralogical Society of America Short Course Notes, v. 6, p. 211–294.

Kastner, M., 1986, New insights into origin of dolomite: 12th International Sedimentologists Congress, Abstracts Volume, p. 158.

Kelts, K., and Hsü, K. J., 1978, Freshwater carbonate sedimentation, in Lerman, A., ed., Lakes: Chemistry, geology, physics: New York, Springer-Verlag, p. 295–324.

Kendall, A. C., 1984, Evaporites, *in* Walker, R. G., ed., Facies models: Geoscience Canada Reprint Series 1, p. 259–296.

Krauskopf, K. B., 1979, Introduction to Geochemistry (second edition): New York, McGraw-Hill Book Company, 615 p.

Lambert, S., 1989, Hydrogeochemistry of saline lakes in the northern Interior Plains of western Canada and northern United States [B.Sc. thesis]: Winnipeg, University of Manitoba, 162 p.

Last, W. M., 1982, Holocene carbonate sedimentation in Lake Manitoba, Canada: Sedimentology, v. 29, p. 691–704.

Last, W. M., 1984, Sedimentology of playa lakes of the northern Great Plains: Canadian Journal of Earth Science, v. 21, p. 107–125.

Last, W. M., 1988, Salt lakes of western Canada: A spatial and temporal geochemical perspective, *in* Nicholaichuk, W., and Steppuhn, H., eds., Proceedings, Symposium on Water Management Affecting the Wet-to-Dry Transition: Saskatoon, Saskatchewan, Water Studies Institute, p. 99–113.

Last, W. M., 1989a, Continental brines and evaporites of the northern Great Plains of Canada: Sedimentary Geology, v. 64, p. 207–221.

Last, W. M., 1989b, Sedimentology of a saline playa in the northern Great Plains, Canada: Sedimentology, v. 36, p. 109–123.

Last, W. M., 1990, Paleochemistry and paleohydrology of Ceylon Lake, a salt-dominated playa basin in the northern Great Plains, Canada: Journal of Paleolimnology, v. 4, p. 219–238.

Last, W. M., and Sauchyn, D. J., 1993, Mineralogy and lithostratigraphy of Harris Lake, southwestern Saskatchewan, Canada: Journal of Paleolimnology, v. 9, p. 23–39.

Last, W. M., and Schweyen, T. H., 1983, Sedimentology and geochemistry of saline lakes of the northern Great Plains: Hydrobiologia, v. 105, p. 245–263.

Last, W. M., and Schweyen, T. H., 1985, Late Holocene history of Waldsea Lake, Saskatchewan, Canada: Quaternary Research, v. 24, p. 219–234.

Last, W. M., and Slezak, L. A., 1987, Sodium sulfate deposits of western Canada, *in* Gilboy, C. F., and Vigrass, L. W., eds., Economic minerals of Saskatchewan: Saskatchewan Geological Society Special Publication 8, p. 197–205.

Lennox, D. H., Maathuis, H., and Pederson, D., 1988, Region 1. Western glaciated plains, *in* Back, W., Rosenshein, J. S., and Seaber, P. R., eds., Hydrogeology: The Geology of North America, v. O-2, p. 115–128.

Lieffers, V. J., and Shay, J. M., 1983, Ephemeral saline lakes in the Canadian prairies—Their classification and management for emergent macrophyte growth: Hydrobiologia, v. 105, p. 85–94.

Löffler, H., ed., 1987, Paleolimnology IV: Hydrobiologia, v. 143, p. 1–431.

Meriläinen, J., Huttunen, P., and Battarbee, R. W., eds., 1983, Paleolimnology: Hydrobiologia, v. 103, p. 1–318.

Morrow, D. W., 1982, Diagenesis, 1. Dolomite—Part 1: The chemistry of dolomitization and dolomite precipitation: Geoscience Canada, v. 9, p. 95–107.

Müller, G., and Wagner, F., 1978, Holocene carbonate evolution in Lake Balaton (Hungary), a response to climate and impact of man, in Matter, A., and Tucker, M. E., eds., Modern and ancient lake sediments: International Association of Sedimentologists Special Publication 2, p. 55–80.

Müller, G., Irion, G., and Förstner, U., 1972, Formation and diagenesis of inorganic Ca-Mg-carbonates in the lacustrine environment: Naturwissenschaften, v. 59, p. 158–164.

Palliser, J., 1862, Journals, detailed reports and observations relative to the exploration by Captain Palliser of a portion of British North America; Exploration of British North America: Report presented to both houses of Parliament, London, 325 p.

Richardson, J. L., and Arndt, J. L., 1989, What use prairie potholes?: Journal of Soil and Water Conservation, v. 44, p. 196–198.

Rawson, D. S., and Moore, G. E., 1944, The saline lakes of Saskatchewan: Canadian Journal of Research (Series D), v. 22, p. 141–201.

Rosen, M. R., Miser, D. E., and Warren, J. K., 1988, Sedimentology, mineralogy and isotopic analysis of Pellet Lake, Coorong Region, South Australia: Sedimentology, v. 35, p. 105–122.

Rozkowska, A. D., and Roskowski, A., 1969, Seasonal changes of slough and lake water chemistry in southern Saskatchewan, Canada: Journal of Hydrology, v. 7, p. 1–13.

Rueffel, P. G., 1968, Development of the largest sodium sulfate deposit in Canada: Canadian Mining and Metallurgical Bulletin, v. 61, p. 1217–1228.

Rust, B. R., and Nanson, G. C., 1989, Bedload transport of mud as pedogenic aggregates in modern and ancient rivers: Sedimentology, v. 36, p. 291–306.

Rutherford, A. A., 1970, Water quality survey of Saskatchewan surface waters: Saskatchewan Research Council, C 70-1, 133 p.

Sahinen, U. M., 1948, Preliminary report on sodium sulfate in Montana: Montana School of Mines Bureau of Mines and Geology Report, 9 p.

Schweger, C. E., and Hickman, M., 1989, Holocene paleohydrology of central Alberta: Testing the general-circulation-model climate simulations: Canadian Journal of Earth Science, v. 26, p. 1826–1833.

Slezak, L. A., and Last, W. M., 1985, Geology of sodium sulfate deposits of the northern Great Plains, *in* Glaser, J. D., and Edwards, J., eds., Twentieth Forum on the Geology of Industrial Minerals: Maryland Geological Survey Special Publication 2, p. 105–115.

Sly, P. G., 1978, Sedimentary processes in lakes, *in* Lerman, A., ed., Lakes: Chemistry, geology, physics: New York, Springer-Verlag, p. 65–90.

Smol, J. P., 1989, Paleolimnology—Recent advances and future challenges, *in* de Bernardi, R., Giussani, G., and Barbanti, L., eds., Scientific perspectives in theoretical and applied limnology: Memorie dell'Instituto Italiano di Idrobiologia, v. 47, p. 247–272.

Smoot, J. P., and Lowenstein, T. K., 1991, Depositional environments of nonmarine evaporites, *in* Melvin, J. L., ed., Evaporites, petroleum and mineral resources: New York, Elsevier, Developments in Sedimentology 50, p. 189–347.

Sonnenfeld, P., 1984, Brines and evaporites: New York, Academic Press, 613 p.

Street-Perrott, F. A., and Harrison, S. P., 1985, Lake levels and climate reconstruction, *in* Hecht, A. D., ed., Paleoclimate analysis and modeling: New York, John Wiley and Sons, p. 291–340.

Talbot, M. R., 1990, A review of the palaeohydrological interpretation of carbon and oxygen ratios in primary lacustrine carbonates: Chemical Geology (Isotope Geoscience Section), v. 80, p. 261–279.

Talbot, M. R., and Kelts, K., 1986, Primary and diagenetic carbonates in the anoxic sediments of Lake Bosumtwi, Ghana: Geology, v. 14, p. 912–916.

Talbot, M. R., and Kelts, K., 1990, Paleolimnological signatures from carbon and oxygen isotopic ratios in carbonates from organic carbon-rich lacustrine sediments, *in* Katz, B. J., ed., Lacustrine basin exploration—Case studies and modern analogs: American Association of Petroleum Geologists Memoir 50, p. 99–109.

Teller, J. T., and Last, W. M., 1981, Late Quaternary history of Lake Manitoba, Canada: Quaternary Research, v. 16, p. 97–116.

Teller, J. T., and Last, W. M., 1990, Paleohydrological indicators in playas and salt lakes, with examples from Canada, Australia, and Africa: Palaeogeography, Palaeoclimatology, Palaeoecology, v. 76, p. 215–240.

Tomkins, R. V., 1954, Magnesium in Saskatchewan: Saskatchewan Department of Mineral Resources Report 11, 23 p.

Vance, R. E., Clague, J. J., Mathewes, R. W., 1993, Holocene paleohydrology of a hypersaline lake in southeastern Alberta: Journal of Paleolimnology, v. 8, p. 103–120.

Wallick, E. I., 1981, Chemical evolution of groundwater in a drainage basin of Holocene age, east-central Alberta, Canada: Journal of Hydrology, v. 54, p. 245–283.

Wallick, E. I., and Krouse, H. R., 1977, Sulfur isotope geochemistry of a groundwater-generated Na_2SO_4/Na_2CO_3 deposit and the associated drainage basin of Horseshoe Lake, Metiskow, east-central Alberta, Canada: 2nd International Symposium on Water-rock Interaction, Strasbourg, France, p. 1156–1164.

Waite, D. T., ed., 1980, Prairie surface waters: Problems and solutions: Canadian Plains Proceedings, v. 7, 102 p.

Waite, D. T., ed., 1986, Evaluating saline waters in a plains environment: Canadian Plains Proceedings, v. 17, 107 p.

Warren, J. K., 1989, Evaporite sedimentology: Importance in hydrocarbon accumulation: Englewood Cliffs, New Jersey, Prentice-Hall, Inc., 285 p.

Winter, T. C., 1977, Classification of the hydrologic settings of lakes in the north central United States: Water Resources Research, v. 13, p. 753–767.

Witkind, I. J., 1952, The localization of sodium sulfate deposits in northeastern Montana and northwestern North Dakota: American Journal of Science, v. 250, p. 667–676.

PALLISER TRIANGLE GLOBAL CHANGE CONTRIBUTION NUMBER 4
MANUSCRIPT ACCEPTED BY THE SOCIETY JULY 2, 1993

Printed in U.S.A.

Geological Society of America
Special Paper 289
1994

Groundwater-discharge playas of the Mallee region, Murray Basin, southeast Australia

Gerry Jacobson, James Ferguson, and W. Ray Evans
Australian Geological Survey Organisation, Box 378, Canberra, A.C.T., 2601 Australia

ABSTRACT

Groundwater discharge complexes, locally known as boinkas, are a distinctive feature of the Mallee region, the western sector of the Murray Basin (300,000 km^2) in southeast Australia. The discharge complexes are nested, with modern playas (salinas) set in late Pleistocene playa sediments and associated aeolian deposits. Playa evolution started with the onset of aridity in the mid-Pleistocene. In the Murray Basin, groundwater discharges, and has discharged in the past, to topographically low parts of the landscape,, including the Murray River itself. The fundamental control on the distribution of the discharge complexes is the presence of subsurface permeability barriers, which are tectonically or stratigraphically controlled, and which disrupt lateral groundwater flow. With the general absence of surface drainage in the Mallee region, the location of the topographic lows is probably influenced by differential compaction of sediments.

Within the modern salinas, brine is generated by evaporative concentration and resolution of salts. The structure of some of the investigated brine pools suggests that density instability leads to brine fingering, sinking, and mixing with regional groundwaters. This process is facilitated by more permeable sediments. "Fossil" brines generated by the late Pleistocene playas can be identified and have contributed to the salinity of the regional unconfined aquifer. Brine pools have migrated in time and space in response to climatic change. This has practical implications for the long-term containment of saline wastewaters, which is a serious problem in the Murray Basin.

INTRODUCTION

The Murray Basin is a Cenozoic sedimentary basin, extending over 300,000 km^2 in southeast Australia (Fig. 1). It is an important agricultural region but suffers from a serious and growing problem of land and water salinization caused by rising water tables in the basin. This is consequent on the clearing of native vegetation for dryland agriculture in the recharge zones of the basin over the last 150 years of European settlement, and also on the irrigation of poorly drained soils. The Murray River Australia's largest river in terms of flow, drains the basin and is affected by increased salinity, to the detriment of important urban and rural water supplies along its course.

The Murray Basin forms a closed groundwater system. The aquifers are recharged by streamflow at the basin margins and by direct infiltration of rainfall. Groundwater discharges to the Murray River and to playas, many of which have been reactivated by rising water tables. The distribution of these features is shown in Figure 2.

The control of rising saline water-tables in irrigation districts is likely to involve the pumping and disposal of saline wastewater for the foreseeable future (Evans, 1989). Playas in groundwater-discharge zones are favored sites for disposal of these saline wastewaters using natural processes of evaporative concentration. There are a growing number of evaporation basins sited or proposed in groundwater discharge zones.

Jacobson, G., Ferguson, J., and Evans, W. R., 1994, Groundwater-discharge playas of the Mallee region, Murray Basin, southeast Australia, *in* Rosen, M. R., ed., Paleoclimate and Basin Evolution of Playa Systems: Boulder, Colorado, Geological Society of America Special Paper 289.

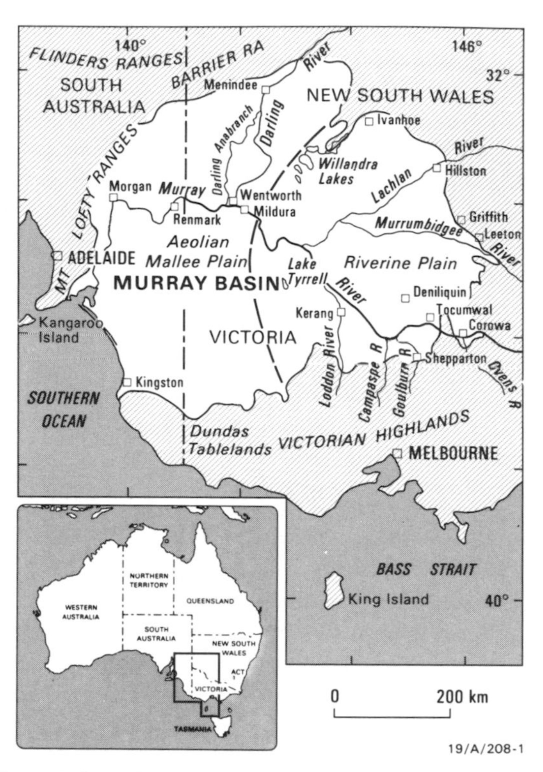

Figure 1. Location map Murray Basin, southeast Australia.

However, the resultant concentrated brines impact on the regional groundwaters and this process is poorly understood. Natural discharge zones vary in their ability to evaporate regional groundwater, and to retain or confine the resulting brines. There is consequently a need for a better understanding of the dynamics of salt accumulation and transport in natural discharge zones, and for a predictive modelling capacity for the impact of saline wastewater disposal.

The importance of groundwater discharge in the development of Australian playas was recognized by P. G. Macumber (1980) and J. M. Bowler (1980, 1986). Macumber (1980) described the role of groundwater discharge in forming distinctive landforms, termed "boinkas", which are nested complexes of salinas and associated dunes and plains, in the Victorian Mallee region of the Murray Basin. The term "Mallee" describes the characteristic vegetation of this semiarid region: eucalypts with a multistemmed habit. Bowler (1986) classified Australian playa basins across climatic zones from humid to arid, and from surface-water dominated to groundwater dominated. Playas with a dominantly surface water regime are characterized by their smooth wave-sculpted outlines, and often have crescentic transverse dunes, called lunettes, on the downwind side. Playas with a dominantly groundwater regime have characteristically irregular outlines, but also commonly have lunettes on the downwind side. The lunettes are composed of pelletal clay, gypsum, or quartz sand, reflecting the lake bed conditions at the time of deflation. The playa brines and the chemistry of their sediments reflect a precipitation series from carbonates or ferric oxides and hydroxides to sulphates to chlorides with increasing solute concentration.

Further investigations by Macumber (1983, 1991) showed that Lake Tyrrell, the largest playa in the southern part of the Murray Basin, has accumulated salt for about 32,000 years, and that hypersaline brine refluxes to the underlying regional aquifer, the Loxton-Parilla Sands (Pliocene Sand), which is saline.

Elsewhere in Australia, Bowler (1986) has demonstrated that the brine pool at Lake Frome, a large playa in South Australia, is about 60 m thick and apparently overlies less saline groundwaters. More detailed work by Allison and Barnes (1985) at Lake Frome identified reverse salinity gradients with hypersaline porewaters at the playa surface becoming appreciably less saline within a depth of 1 m. At Lake Amadeus, a large playa in central Australia, both the stable condition of brine becoming more saline with depth, and the unstable condition of brine becoming less saline with depth, occur (Jacobson, 1988). Recent drilling beneath the temporarily dry bed of Lake George in southeast Australia, shows increasing porewater salinity to a depth of 10 to 12 m in clay beneath the lake bed, then decreasing salinity below that to a depth of 50 m (Jacobson et al., 1991).

Evaporation of water from saturated, saline sediments in playas results in increased density, and concentration of conservative solutes such as chloride near the playa surface. The density difference may reverse the upwards hydraulic gradient, allowing the possibility of advective reflux of brines to the underlying aquifer. The chloride concentration difference can induce a vertically downwards diffusive flux of salt even against the upwards hydraulic gradient. If the evaporation rate and sediment permeability are high, then advection is likely to be more effective than diffusion in transporting salt downwards (Barnes and others, 1991). Theoretical and laboratory studies by Wooding (1969, 1989) have shown that the density instability may produce small-scale fingering of brine. Fingers of dense hypersaline brine may sink to considerable depths, mixing advectively with the deeper groundwater. The brine fingering process may explain the brine pool stratification observed at several Australian playas, although the process has not yet been observed in nature.

In this paper we review the regional hydrogeological setting of groundwater-discharge complexes in the western Murray Basin and report preliminary investigations of playas in these complexes. These investigations have been undertaken in order to test the salt accumulation and transport hypothesis, and quantify parameters for modelling the process.

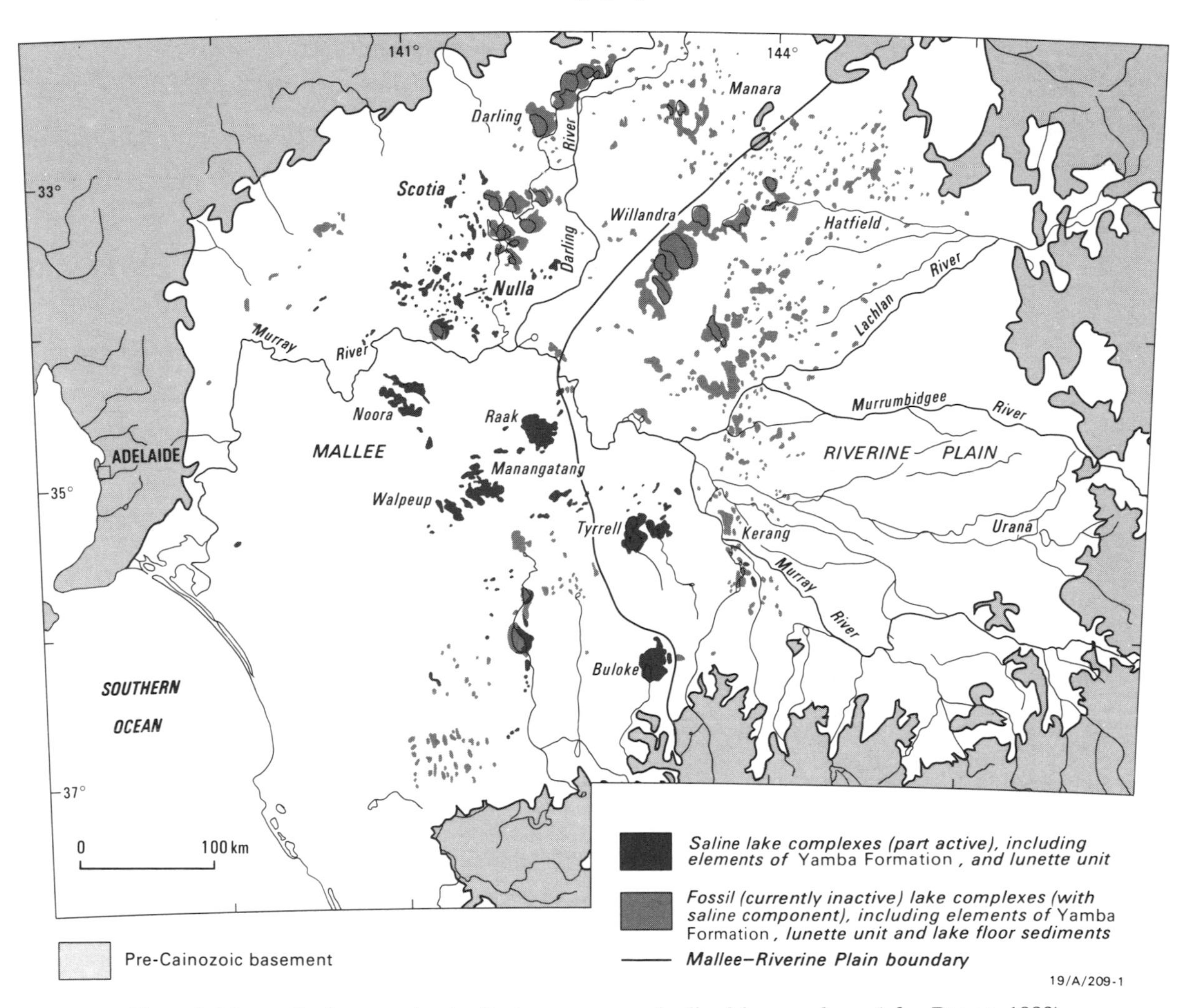

Figure 2. Murray Basin: groundwater discharge zones and saline lake complexes (after Brown, 1989).

REGIONAL GEOLOGY

The Murray Basin is underlain by folded and block-faulted rocks of Proterozoic and lower Palaeozoic age (Brown and Stephenson, 1991). The basin contains Cenozoic sediments which are as much as 600 m thick in the western depocenter, the Mallee region, and 400 m thick in the eastern depocenter, the western Riverine Plain. Elsewhere in the basin, the Cenozoic sediments are generally 200 to 300 m thick, forming a thin, platform-cover succession.

The Tertiary stratigraphy of the Murray Basin is shown schematically in Figure 3. There are three main depositional sequences, of Paleocene to early Oligocene (mainly Eocene), late Oligocene to middle Miocene, and late Miocene to Pliocene age, respectively (Brown, 1989). Each of these depositional sequences comprises several genetically related formations separated by disconformities, which reflect transgressive-regressive marine cycles. The early Tertiary sequence is predominantly nonmarine throughout the basin. The younger two sequences contain marine sediments in the central and southwest sectors of the basin, but mainly nonmarine sand, silt, clay, and carbonaceous sediments in the east and north.

In the east of the basin, the Tertiary sediments are overlain by Quaternary fluvio-lacustrine and aeolian sediments of the Riverine Plain. The west, the Mallee region, was partially covered by a megalake, Lake Bungunnia, in the late Pliocene and early Pleistocene (Stephenson, 1986). About 20 m of clay (Blanchetown Clay) was deposited at that time, in a climate wetter than at present. The Blanchetown Clay overlies Pliocene sand and was partly deposited between Pliocene beach ridges.

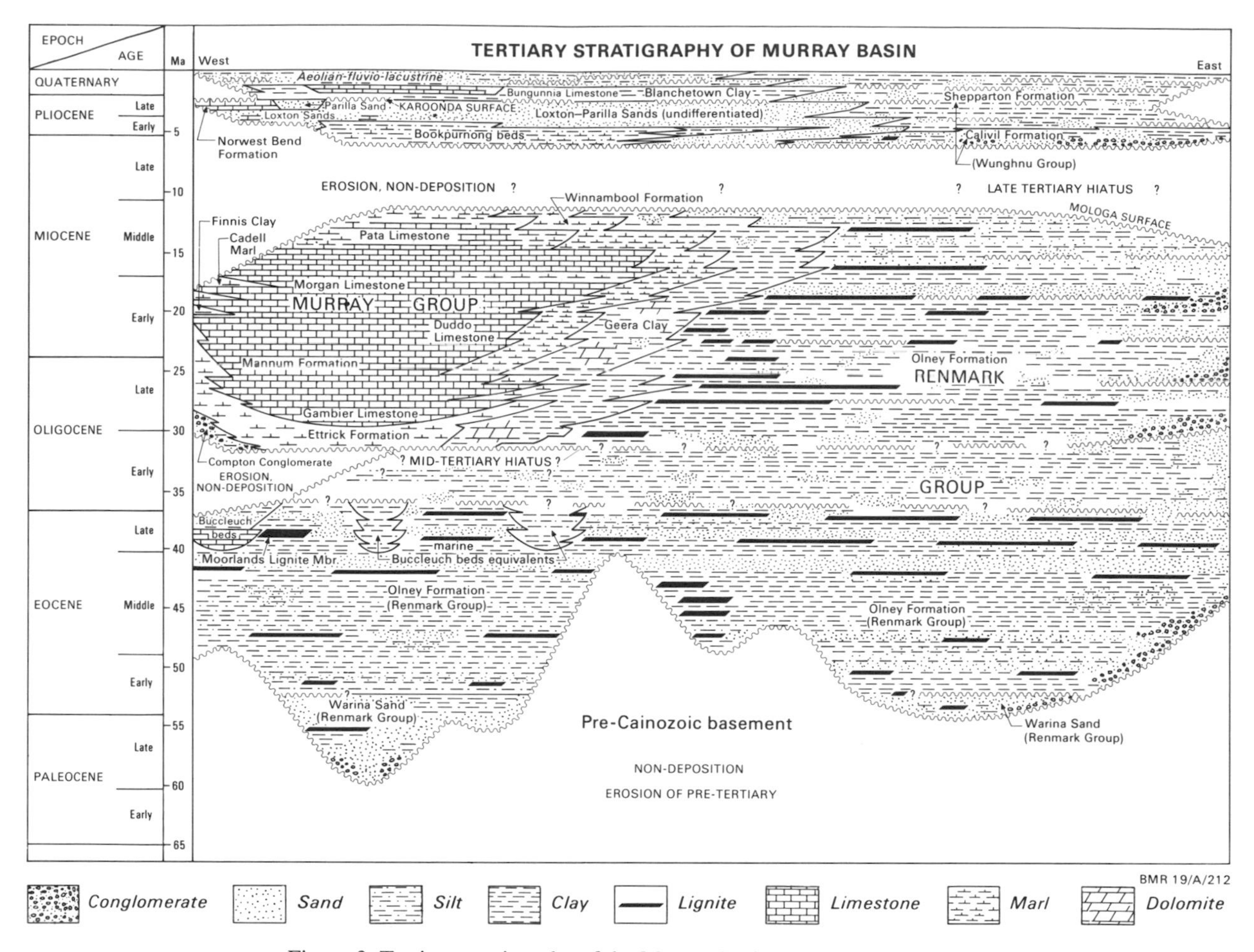

Figure 3. Tertiary stratigraphy of the Murray Basin (after Brown, 1989).

Drying of this lake between 0.7 and 0.4 Ma began a period of generally arid climatic conditions lasting till the present day. In this period of time, groundwater discharge complexes have developed in topographically low areas, some of which expose the Blanchetown Clay (Brown, 1989; Stephenson, 1986).

In the Mallee region the Tertiary and early Pleistocene sediments are generally covered by late Pleistocene aeolian dunefields, with minor fluvial and lacustrine sediments. The Mallee-Riverine boundary (Fig. 2) is significant in terms of drainage; only the Murray River and the Darling River flow through the Mallee whereas the Riverine Plain is traversed by several rivers.

Continuous small-scale tectonics throughout the Cenozoic has been the major influence on sedimentation patterns (Brown and Stephenson, 1991). Examples of late Tertiary and Quaternary tectonics with significant surface expression include: the formation of Lake Bungunnia by tectonic damming of the pro-to–Murray River at 2.5 Ma (Stephenson, 1986); displacement of Pliocene-Miocene sediments by 30 m since 1.7 Ma along the Tyrrell Warp (An et al., 1986); and change in the course of the Murray River at Echuca by late Quaternary movement on the Cadell fault (Bowler and Harford, 1966).

A late Pleistocene time scale for hydrological change has been elucidated by Bowler (1980) for the Willandra Lakes in the northeast Mallee (Fig. 2), and Bowler and Teller (1986) for Lake Tyrrell in the southeast Mallee. From about 55,000 to about 36,000 yr B.P. the climate was wetter than at present (Bowler, 1980). During this period, the "Mungo Lacustral Phase", there was fresh water in the Willandra Lakes, and deep water at Lake Tyrrell. Drier conditions after about 36,000 yr B.P. led to lower lake levels. Thus Lake Tyrrell was dry at about 28,000 yr B.P. and again at about 22,000 yr B.P. (Bowler and Teller, 1986). A dry period at the Willandra Lakes coincided with maximum dune building at about 16,000

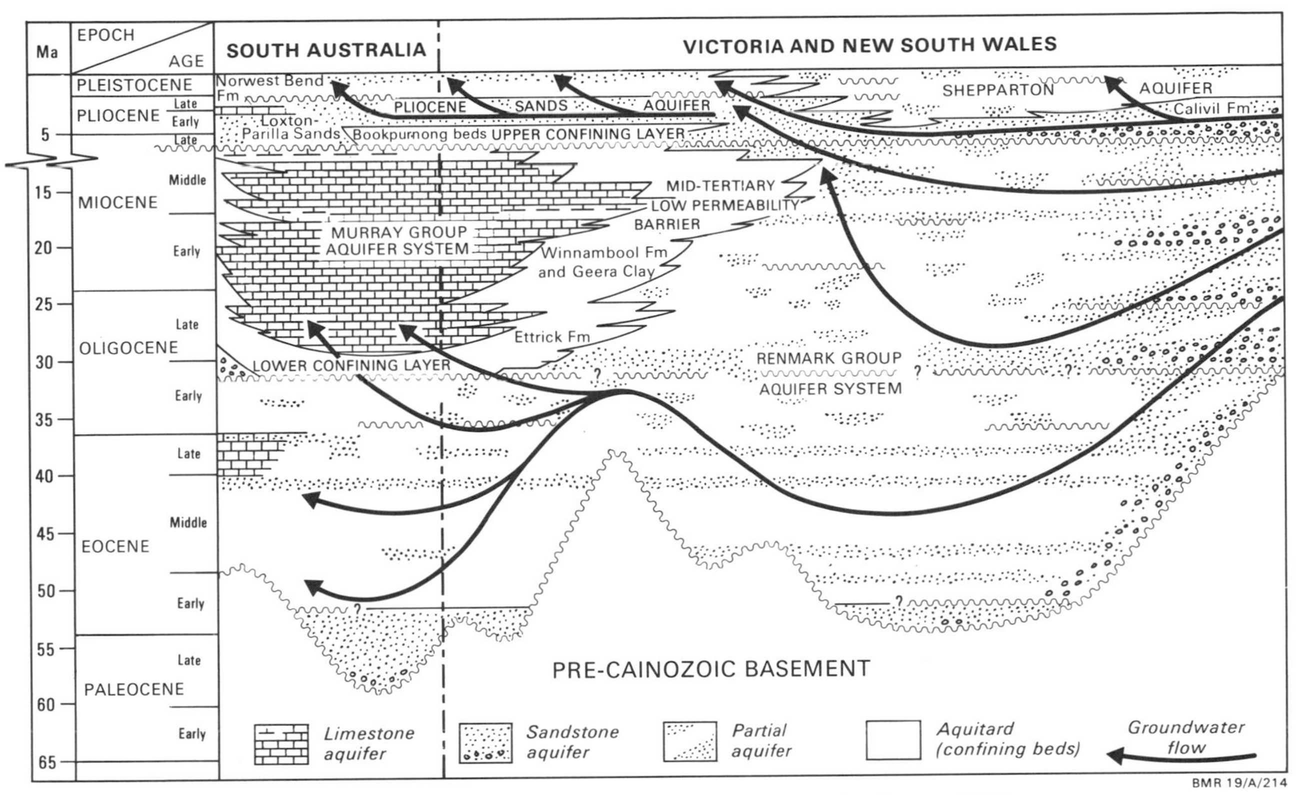

Figure 4. Murray Basin: stratigraphy and aquifer systems (after Brown, 1989).

yr B.P., the "Zanci Phase" (Bowler, 1980). Following this, the period from 10,000 to 6,000 yr B.P. was probably slightly wetter than at present.

REGIONAL HYDROGEOLOGY

Aquifer/aquitard relationships in the Murray Basin are shown schematically in Figure 4. The main fluvial sand unit of the Palaeocene-lower Oligocene sequence forms the Renmark Group, which is as much as 300 m thick. The upper Oligocene–middle Miocene sequence contains shallow-marine limestones collectively known as the Murray Group, and marginal-marine clays and marls, the Geera Clay and Winnambool Formation. In the eastern part of the basin, the Renmark Group extends up into the Miocene and partly envelops the Geera Clay. The upper Miocene–Pliocene sequence contains an important Pliocene marine sand unit, the Loxton-Parilla Sands, in the west, and a widespread Pliocene fluviatile unit, the Calivil Formation, in the east. These units are overlain, in the east of the basin, by the Shepparton Formation: clay, silt, and sand, of Pliocene to Quaternary age.

The main aquifers are the Renmark Group, Murray Group limestone, Pliocene Sand, which is a composite of the Loxton-Parilla Sands and Calivil Formation, and Shepparton Formation (Evans et al., 1990). The Renmark Group forms the basal aquifer throughout the basin. It is recharged toward the basin margin, and groundwater flow is inwards, towards the west-central depocenter of the basin, where it leaks upwards into the Murray Group aquifer.

The Murray Group limestone aquifer is best developed south of the River Murray where it is an important freshwater source. The travel time along a 300-km flowline is estimated as 150,000 years (Davie et al., 1989). The aquifer is more saline immediately north of the river, where it may be partly recharged by upwards leakage from the Renmark Group. Murray Group groundwater flows towards the river, where its discharge is the main cause of significantly increasing river salinity in South Australia.

The Pliocene Sand aquifer is widespread in the basin (Fig. 5); it is generally unconfined in the west, but is partly confined by Quaternary units in the east. It is recharged around the basin margin by infiltrating rainfall where the aquifer is unconfined, by downwards leakage from the overlying Shepparton Formation, and by channel bed leakage from streams in the Riverine Plain. The aquifer is fresh at the eastern, northeast, and southeast margins but saline elsewhere, and

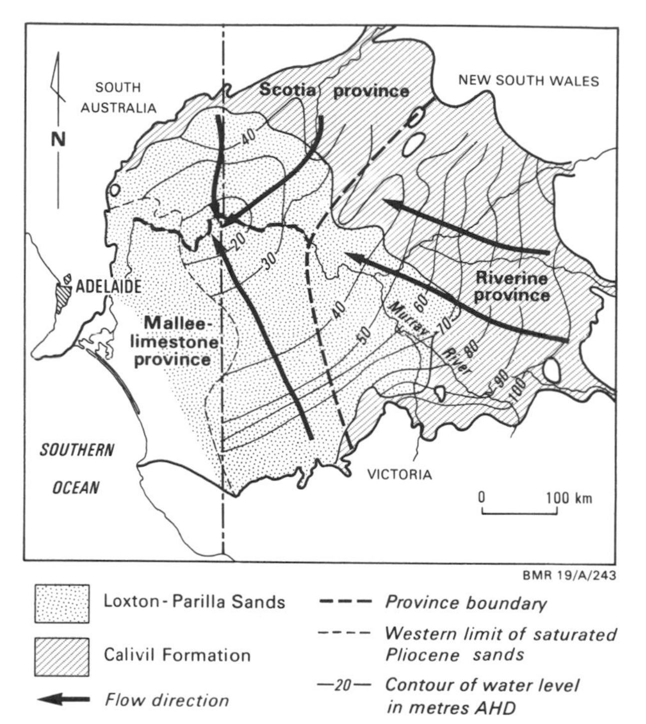

Figure 5. Groundwater flow systems in the Pliocene Sand aquifer, Murray Basin (after Evans and Kellett, 1989).

underlies the discharge complexes described in this paper. The Pliocene Sand aquifer in the west of the basin has a similar chemical composition to that of seawater in many respects (Macumber, 1983, 1991).

The main aquitard in the Murray Basin is a composite of mid-Tertiary low-permeability clays and marls (Winnambool Formation, Geera Clay) that extend in an arc about 100 km wide through the center of the basin (Evans and Kellett, 1989). This low-permeability barrier disrupts lateral groundwater flow in the Renmark Group, and controls the distribution of groundwater discharge zones (Brown and Radke, 1989). In the west and south, the Renmark Group is separated from the Murray Group limestone by a lower confining layer, the Ettrick Formation (Evans et al., 1990). The eastern part of the Murray Group limestone is separated from the Pliocene Sand by an upper confining layer, the Bookpurnong Beds.

The occurrence of groundwater discharge zones in the Murray Basin is influenced by disruption of groundwater flow and upwards movement at subsurface permeability barriers. These barriers occur where aquifers are thinned by rising basement or lateral changes in lithology. Thus, groundwater discharge into the Murray River in South Australia is known to be structurally controlled (Telfer, 1991; Lindsay and Barnett, 1989).

The regional aquifers of the Murray Basin can be divided areally into three distinct provinces (Fig. 5): the Riverine province in the east, the Mallee-Limestone province in the southwest, and the Scotia province in the northwest (Evans and Kellett, 1989). The two groundwater discharge complexes described in the present study, are situated in the Scotia groundwater province in western New South Wales (Fig. 5). In this province, groundwater flows from north to south, in the shallow unconfined Pliocene Sand aquifer and in the deeper, confined Renmark Group aquifer. These aquifers are separated by low-permeability units, the Geera Clay, Winnambool Formation, and Bookpurnong Beds. Recharge to both major aquifers is from outwash from ephemeral streams on the northwest basin margin, and recharge to the Pliocene Sand is also by direct infiltration of rainfall. Upwards leakage of Renmark Group groundwater into the Murray Group aquifer occurs in the west (Lindsay and Barnett, 1989; Telfer, 1991). Groundwater discharges to the numerous playas in the region (Fig. 6) and to the Murray River.

The climate in the Mallee region is semiarid. Annual rainfall at Mildura is about 270 to 320 mm, and pan evaporation is about 2,200 mm. Estimates of groundwater recharge range from 0.04 to 0.08 mm per year beneath sand dunes with native mallee vegetation (Allison and Hughes, 1983) to 3 to 10 mm per year beneath cleared and cropped land (Cook and Walker, 1989). The increase in recharge consequent on land clearing in the Mallee region causes local dryland salinity problems. The fundamental reason for the basinwide rise in water tables is the increasing pressure generated by land clearing along the foothills of the Great Dividing Range, which is the main regional recharge zone.

NESTED GROUNDWATER DISCHARGE COMPLEXES

The general character of nested groundwater discharge complexes in the Mallee region of the Murray Basin has been described by Macumber (1980). These features, termed boinkas in northern Victoria, are developed in depressions in the landscape, and have a characteristically irregular outline. The depressions contain a suite of landforms that includes salinas, sand and gypsum flats, and source-bordering dunes. There is generally a distinct outer perimeter that separates these landforms from the surrounding Mallee dune system. The perimeter is commonly formed by lunettes or may be a sharp erosional contact (Figs. 7, 8). The largest boinka in the basin is Raak Plain in northern Victoria, which covers about 550 km^2 (Fig. 2).

In the present investigation, reconnaissance surveys of groundwater-discharge complexes have been carried out in the Mallee region of western New South Wales (Fig. 1). This is a region of no surface drainage, except for the Murray River and the Darling River and its Anabranch, which bound the region.

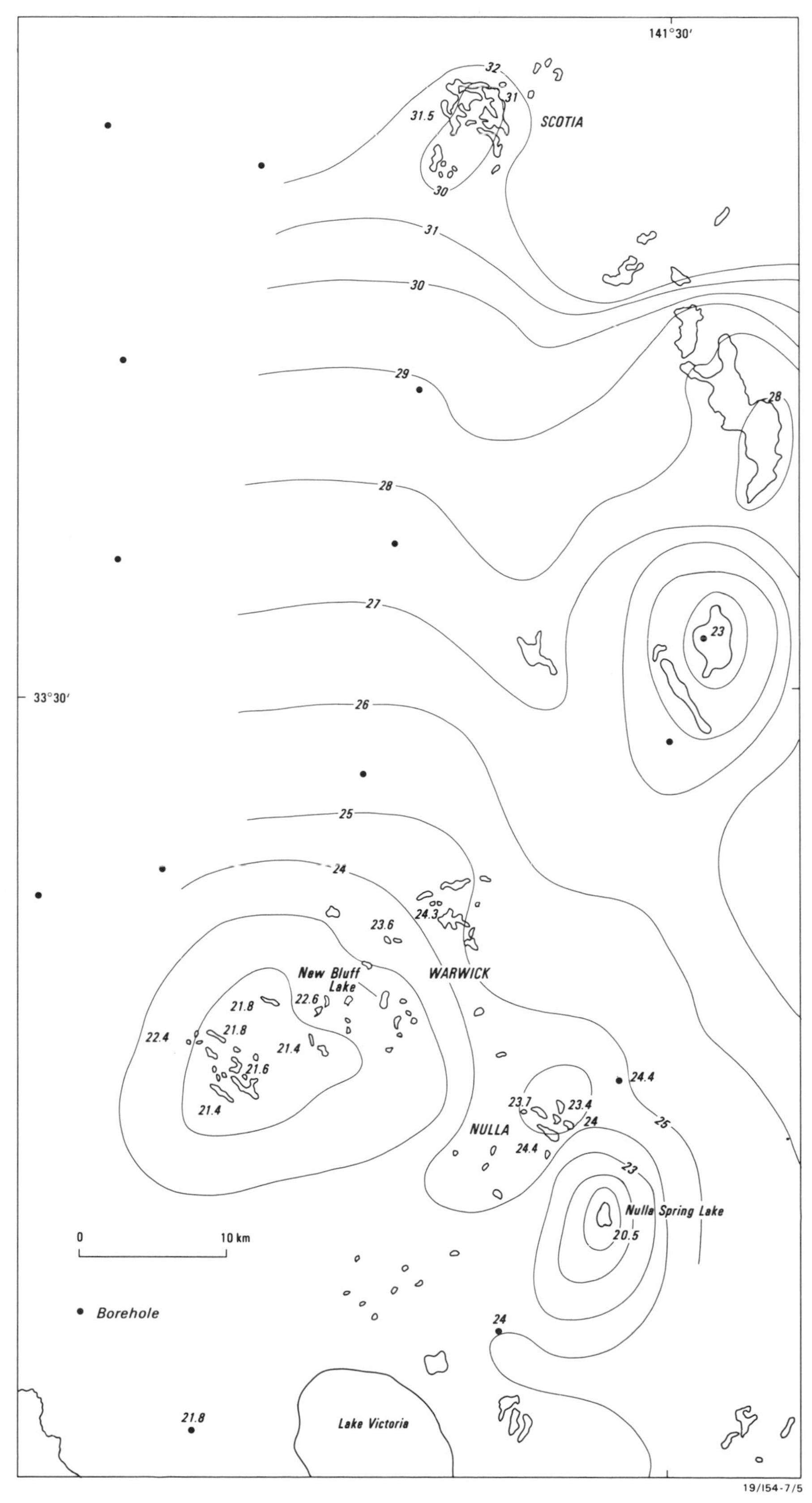

Figure 6. Water level contours in the Pliocene sand aquifer in the Nulla-Scotia region, Murray Basin.

Salinas within the groundwater discharge complexes are topographically low points in the landscape (Fig. 6). Nulla Spring Lake is the lowest, with a bed elevation of 20.5 m above sea level. The groundwater discharge complexes have coalesced in the south of this region to form a distinctive groundwater discharge landscape that contrasts with the surrounding and interspersed Mallee sand dunes.

The topographic lows that form the locus for groundwater discharge may be due to the differential compaction of sediments. It is significant that this region is the western depocenter of the Murray Basin, with more than 400 m of sediments (Brown, 1989). Figure 7 shows the salina, New Bluff Lake, in the Warwick discharge complex; this salina is bounded on the west by a curving escarpment in Blanchetown Clay, sugges-

Figure 7. Aerial photo of New Bluff Lake, showing scarp on the western side. The scarp is about 20 m high, and exposes Blanchetown Clay. The salina is set in an older lacustrine deposit at an elevation of about 25 m. The location is shown in Figure 6.

Figure 8. Aerial photo of the Scotia groundwater discharge complex, set in linear east-west-trending Mallee dune ridges, visible on the west side. Crescentic dunes (lunettes) are visible in the complex, and to the east. White linear features are cleared seismic exploration lines.

tive of recent subsidence. The salina bed level is 20 m below the former bed of Lake Bungunnia. The southern part of the region is underlain by Murray Group limestone, up to 150 m thick, at a depth of about 150 m beneath a cover of Pliocene Sand, Bookpurnong Beds, Geera Clay, and Winnambool Formation. Hills (1940) suggested that distribution of some of the Mallee depressions containing playas, may be due to solution of underlying limestone.

Two salinas in this region have been selected for detailed study of the hydrodynamic processes involved. These are Nulla Spring Lake, which is a salina in the Nulla groundwater discharge complex; and the northeast salina of the Scotia groundwater discharge complex (Fig. 2). These salinas demonstrate end points of a lithological and chemical spectrum. Nulla Spring Lake is representative of a clay-bottomed salina, while the northeast Scotia salina contains significant areas of sand. At Nulla, the lacustrine sediments contain dolomitic and calcitic carbonates, whereas at Scotia the sediments contain ironstone but very little carbonate.

Special technology has been developed for salina drilling by technical staff of the Australian Geological Survey Organisation. This has enabled coring to a depth of 12 m in soft and wet surface conditions. A tripod-mounted drilling rig is assembled on site using a hovercraft to transport the components on the salina. Porewaters are extracted from the drillcores in a field laboratory so that salinity profiles can be determined on site. In addition, rotary rigs have been used to install deeper (to 60 m) piezometers in the regional groundwaters.

SCOTIA GROUNDWATER DISCHARGE COMPLEX

The Scotia Groundwater Discharge Complex occurs in a setting of uncleared Mallee woodland (Fig. 8). Its geology is shown in Figure 9. Ferruginized sandstone (Pliocene Sand) crops out on the north and west sides of the complex where the modern salinas are eroding into upper Pleistocene sand dunes.

The Scotia groundwater discharge complex has similar geomorphological features to those described by Macumber (1980) for Raak Plain and other discharge complexes in the southern Mallee. Thus it is a "nested" complex with modern salinas at an elevation of 31 m above sea level set in older lacustrine and associated aeolian and groundwater deposits. These older deposits may be of late Pleistocene age, equivalent to the Mungo Lacustral Phase (Bowler, 1980), the last known humid phase in this region. The lacustrine sediments comprise several meters of gray clay and form terraces, generally 2 to 3 m above the modern salina level. Near the top of the lacustrine sequence is a bed of groundwater-precipitated

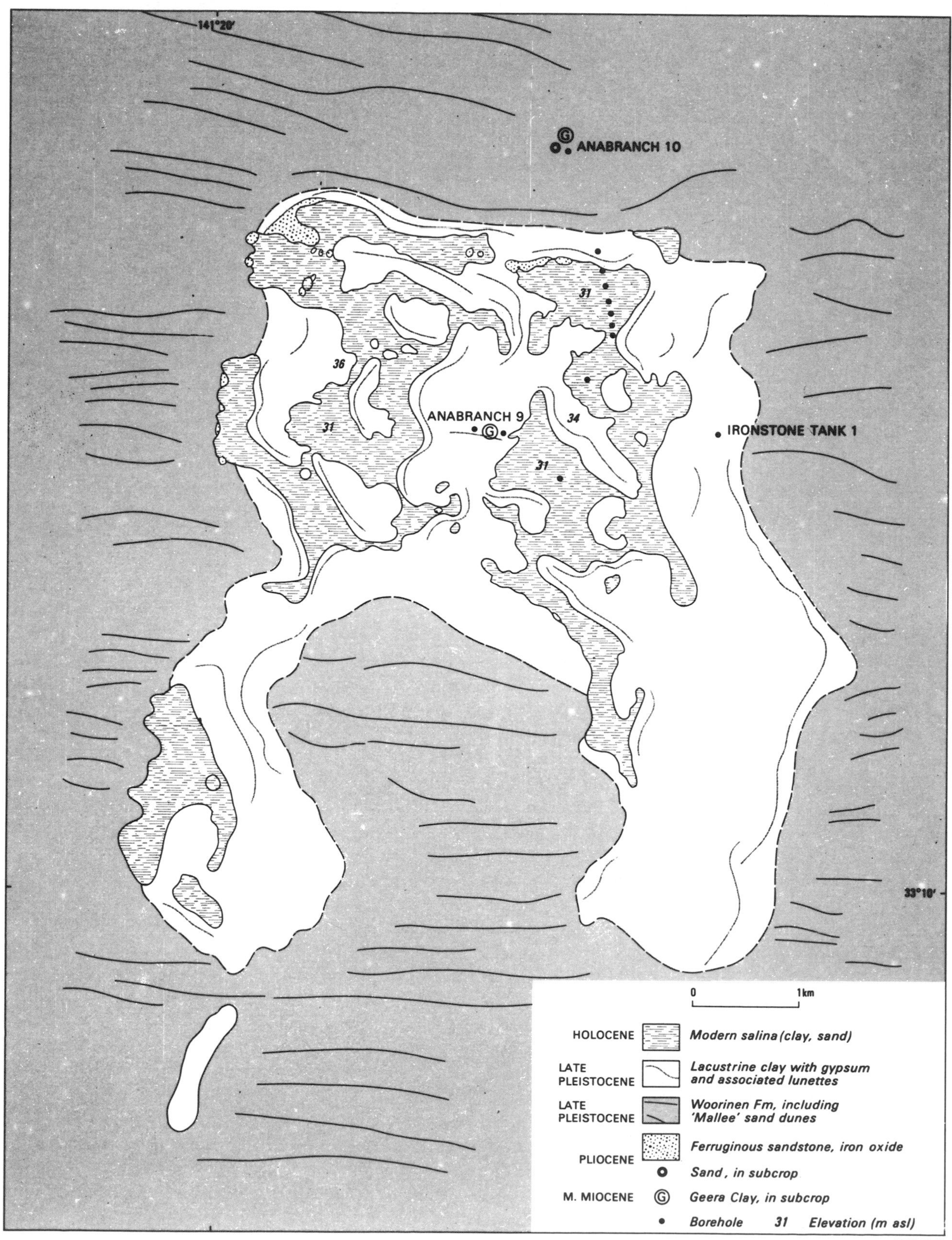

Figure 9. Geological map of the Scotia groundwater discharge complex.

gypsum up to 60 cm thick, representing the drying-out phase of this former lake. Lunettes, crescentic dunes, up to 10 m high, border the former lake sequence. At least two generations of lunettes are apparent on the east side of the complex: an outer sand lunette, and inner gypseous clay lunettes. Ironstone forms terraces up to 2 m above modern salina level on the north and west sides of the complex; this represents former springs that discharged into the late Pleistocene lake.

In the northeast Scotia salina, a transect of drillholes to a depth of about 10 m intersected a thin layer (up to 1 m) of modern lacustrine clay overlying sand with thin clay intercalations, probably the Pliocene Sand (Fig. 10). The sand extends to 6 to 10 m below the lake bed where it overlies Geera Clay. Two investigation drillholes to depths of 60 m have intersected sands of the Renmark Group below the Geera Clay.

Salinity in the lacustrine clay is high, up to 250 g/L total dissolved solids (TDS), and the salinity-depth profiles are complex (Fig. 11). They probably result from a combination of diffusion and the periodic infiltration of surface water containing redissolved salts, into dessication cracks. Groundwater in the underlying sand is considerably more saline (around 140 g/L) than the inflowing regional groundwaters (40 g/L). The highest concentrations occur at the margin of the salina where the surficial clays grade into sands. Profiles show a salinity minimum between the hypersaline brines in the surficial clay and the underlying, more saline regional groundwaters. These saline groundwaters possibly result from advective reflux in the past, before the deposition of the surficial clay that presently inhibits advection.

Beneath the Pliocene Sand, salt has moved through the Geera Clay aquitard, probably by diffusion. The salinity of the underlying Renmark Group aquifer has been enhanced from about 28 g/L just north of Scotia, up to 54 g/L beneath the center of the discharge complex (Fig. 10).

The Pliocene Sand groundwaters near the Scotia discharge complex are neutral (pH 6.9), saline (40 g/L) waters of the Cl-SO_4-Na-Mg type (Table 1). The shallow salina brines are also Cl-SO_4-Na-Mg waters but contain proportionately more $MgSO_4$ and NaCl than would evaporated Pliocene Sand groundwater, suggesting resolution of salt. The salina brines are also considerably more acid, down to pH 4.

The Pliocene Sand groundwaters at Scotia are moderately reducing (Eh, –72 mV) and have significant dissolved iron concentrations, as do the underlying Renmark Group groundwaters. High dissolved iron concentrations are common in Australian continental aquifer systems (Ferguson et al., 1983; Macumber, 1983, 1991). Dissolved iron may be produced by reduction and dissolution of Fe(III) oxides and hydroxides coating sand grains in the aquifer. This reduction is facilitated by fermentative bacteria, which consume organic material in the sediments, remove dissolved oxygen, and produce CO_2. If there is sufficient organic material then bacterial sulphate reduction may also occur, and iron sulphide will precipitate on the sediment grains. Emergence of these groundwaters into the near-surface environment, results in oxidation of dissolved Fe(II) to Fe(III), followed by hydrolysis of the Fe(III), and precipitation of iron oxides, or oxidation of the iron sulphides to produce sulphuric acid and dissolved Fe(III). At Scotia, acid hypersaline brines have been produced as iron-rich groundwaters flow into the salinas. The extent of the Pleistocene ironstone deposits indicates a more iron-rich source during an earlier phase of lake deposition.

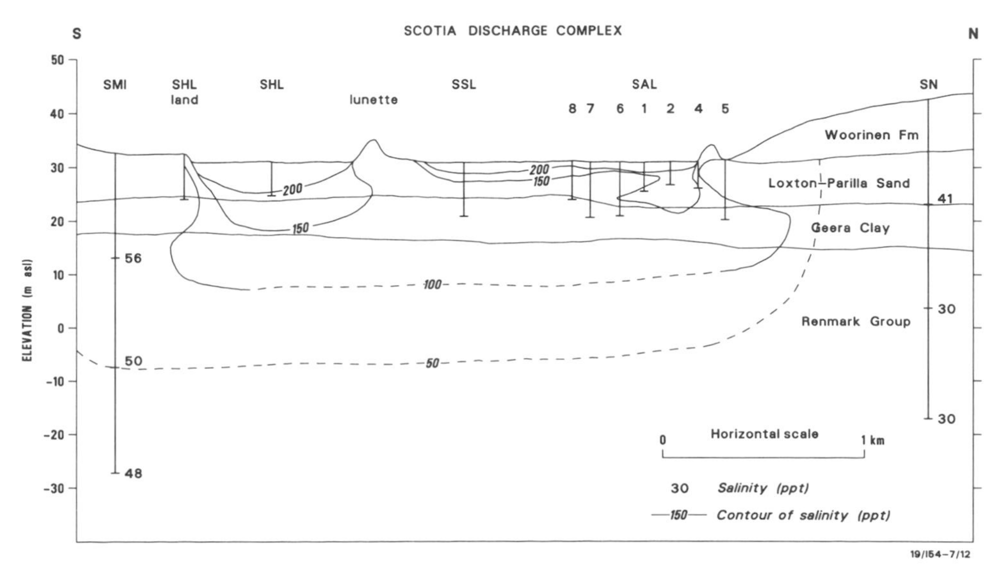

Figure 10. Transect showing brine salinity in the east Scotia salina. Bore marked "SN" is Anabranch 10 (Fig. 9); bore marked "SMI" is Anabranch 9 (Fig. 9).

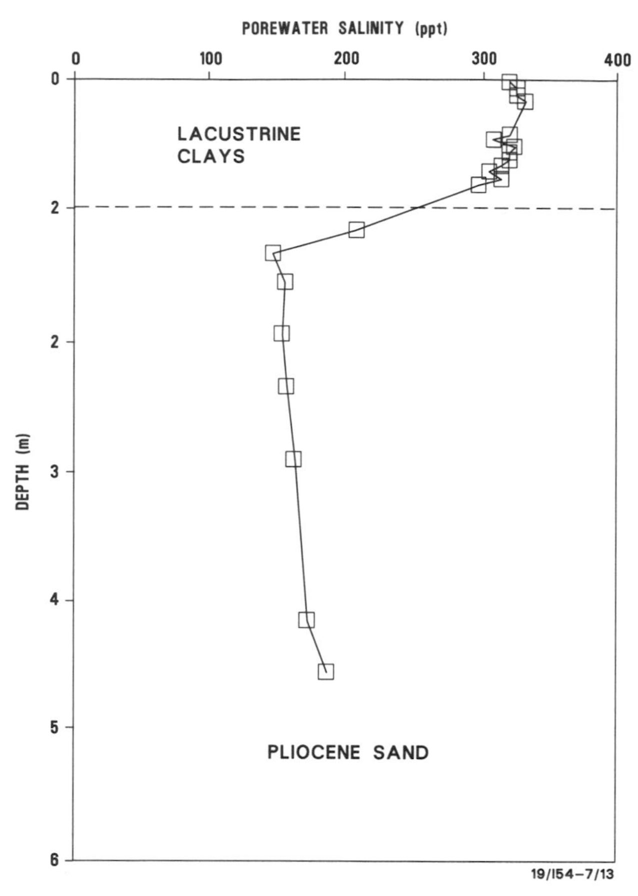

Figure 11. Salinity profile at the northeast Scotia salina.

Groundwater at the top of the Renmark group aquifer is of the Cl-Na-(Mg) type, and is affected by diffusion of salt through the Geera Clay aquitard.

NULLA GROUNDWATER DISCHARGE COMPLEX

The Nulla Groundwater Discharge Complex contains remnants of native Mallee woodland, but is generally saltbush. Its geology is shown in Figure 12.

At Nulla Spring Lake, a transect to a depth of about 10 m across the salina has intersected several meters of upper Pleistocene lacustrine clay, carbonate, and gypsum. Two generations of lunettes are discernible, bordering the modern salinas and the upper Pleistocene lacustrine deposits. The sediments overlie the regional Pliocene Sand aquifer.

The lacustrine clays at Nulla Spring Lake contain thin dolomite beds, and the upper part of the sequence is gypsiferous. The lacustrine sequence probably correlates with wet conditions of the Mungo Lacustral Phase at the Willandra Lakes (Bowler, 1980) and Lake Tyrrell (Bowler and Teller, 1986). The upper Pleistocene lacustrine deposits can be traced throughout the Warwick and Nulla discharge complexes, forming a widespread terrace at an elevation of about 25 m; the modern salinas are cut in at 20 to 21 m. This indicates a loss of groundwater head of about 4 m since the late Pleistocene wet period. This probably reflects climatic change although the location of the active salinas may also be influenced by deflation or subsidence.

Regional groundwater discharges to Nulla Spring Lake from numerous small springs on a north-south lineament. The groundwater evaporates to form hypersaline brines (up to 340 g/L TDS) close to the salina surface. Salinity profiles of pore-

TABLE 1. CHEMICAL ANALYSES OF REGIONAL GROUNDWATERS AND BRINES

	North Scotia (Anabranch 10)	Scotia Salina	Scotia Island (Anabranch 9)	Nulla 9 (Anabranch 8)	Nulla Salina	Nulla 5 (Anabranch 6)
Depth (m)	20	0.12	20	20	0.65	60
Aquifer	Pliocene Sand	Late Quaternary	Renmark Group	Pliocene Sand	Late Quaternary	Pliocene Sand
TDS	40,000	252,000	54,000	26,000	320,000	106,000
pH	6.92	5.88	6.24	7.88	6.75	7.27
Ca	967	486	727	417	391	694
Mg	1,830	20,290	2,320	950	15,200	3,770
Na	11,400	65,575	17,200	8,550	93,100	34,300
K	174	713	248	67	458	229
SiO_2	14	16	1	7	9	3
HCO_3	237	92	55	365	250	372
Br	56	45	58	33	235	98
Cl	18,700	92,600	28,200	14,000	150,000	60,800
SO_4	8,520	79,400	7,050	3,150	46,400	11,700
Fe	5.45	1.04	40.90	0.10	1.10	3.51

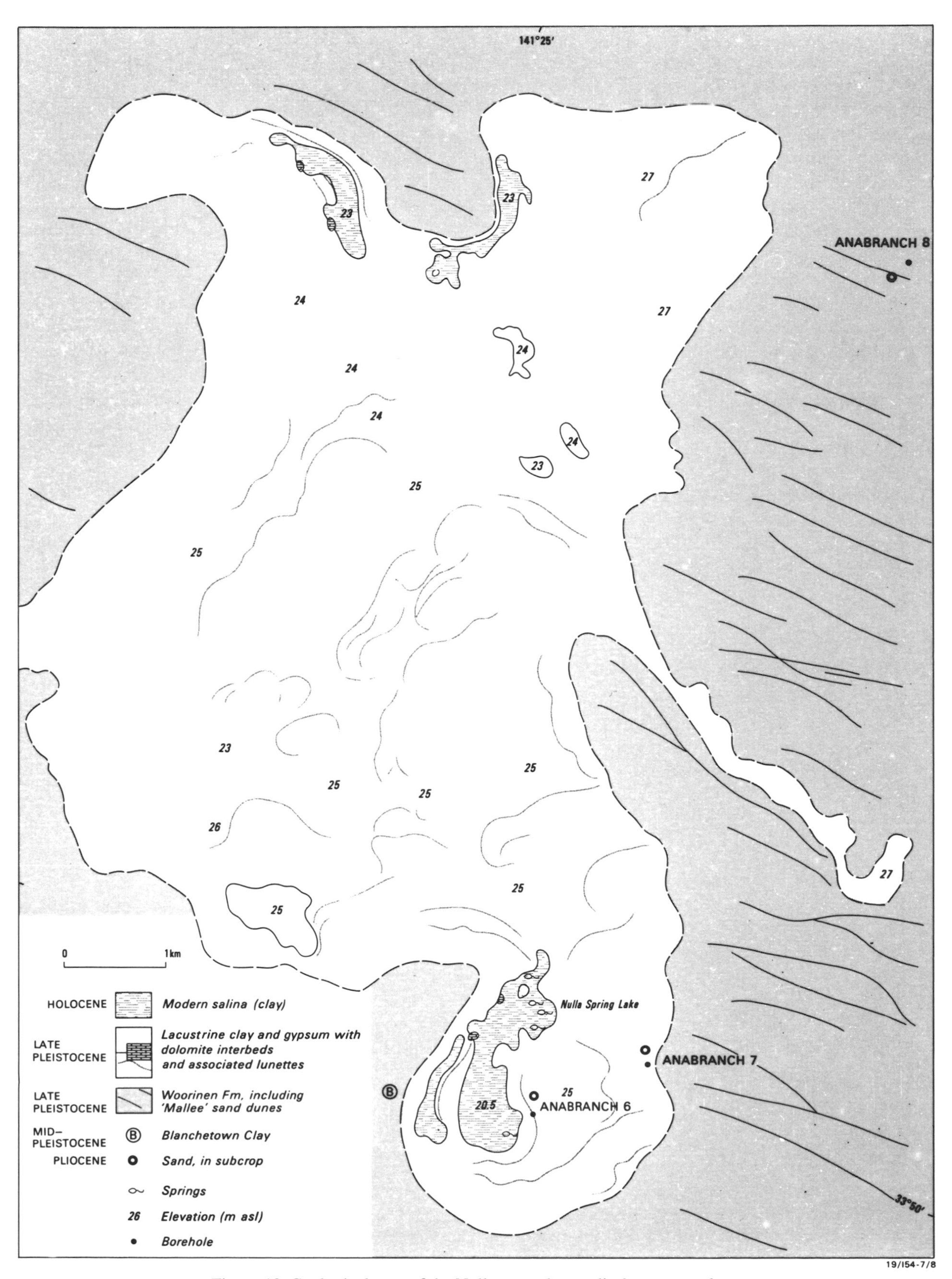

Figure 12. Geological map of the Nulla groundwater discharge complex.

waters (Fig. 13) indicate that the hypersaline brine is concentrated near the surface of the clay and diffuses downwards through the clay to regional groundwater in the underlying Pliocene Sand aquifer (70 g/L TDS). The salinity pattern contrasts to that of the northeast Scotia saline (Fig. 11) where brine has sunk advectively in a sand layer. The onset of hydraulic instability leading to brine fingering and the advective downwards transport of salt is determined by a critical Rayleigh number. The critical Rayleigh number is proportional both to the density difference between the brine layer and the underlying regional groundwater, and to the ratio of the hydraulic conductivity of the lake sediments to the evaporation rate (Wooding, 1989; Barnes et al., 1991). Thus conditions for advective downwards transport of salt appear to have been achieved at the northeast Scotia salina which is mainly underlain by sand, but not at Nulla Spring Lake which is underlain by clay. The evaporation rate is probably similar at these two salinas, which are in the same climatic zone, but the hydraulic conductivity is markedly different.

Extension of the transect in depth and eastwards across the late Pleistocene lacustrine and lunette deposits (Fig. 14), indicates the probable sinking of "fossil" brine generated by the late Pleistocene salina. Regional Pliocene Sand groundwaters near Nulla Spring Lake are of the Cl-Na-(Mg) type (Table 1).

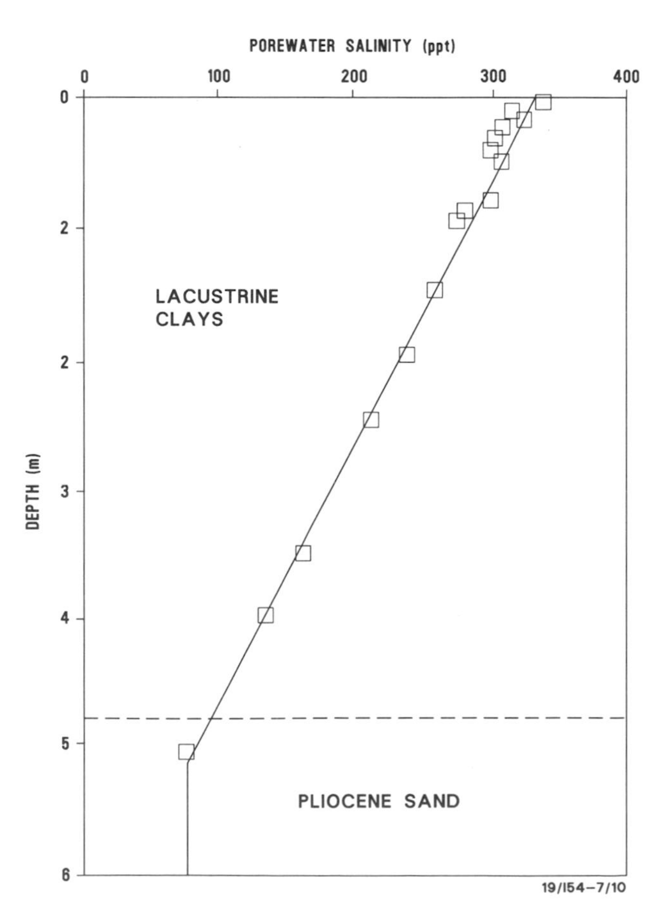

Figure 13. Salinity profile in clay at Nulla Spring Lake (bore NSL 11).

They have high alkalinity, up to 600 mg/L, near neutral pH and high pP_{CO_2}, but low dissolved iron concentrations. This probably reflects biological processes within the aquifer having advanced to the stage where sulphide from bacterial sulphate reduction has precipitated most of the dissolved Fe(II) from solution, and has simultaneously generated alkalinity. As a result, the Pliocene Sand groundwater at Nulla has the potential to precipitate carbonate rather than iron hydroxides, and this is reflected in the composition of the lacustrine sediments. The shallow salina brines are Cl-(SO_4)-Na-Mg waters, whereas the refluxed "fossil" brines are of the Cl-Na(-Mg) type.

Nulla Spring Lake is a natural analogue for an evaporation basin, with its impermeable base and its seasonal filling with surface water from local rainfall. Preliminary results of groundwater monitoring suggest that desired containment of evaporated brines could be achieved at Nulla because of its clay seal. The slow diffusion through the clay indicates that containment would be effective for some thousands of years.

CONCLUSIONS

1. Distinctive nested groundwater-discharge complexes (boinkas) in the western Murray Basin demonstrate playa evolution in and since the late Quaternary.

2. Groundwater discharges at topographically low points in the landscape, as a result of disruption to groundwater flow at subsurface permeability barriers. The distribution of the topographic lows may be influenced by subsidence in the underlying sediment pile.

3. Preliminary investigations at Scotia and Nulla salinas indicate contrasting hydrological and geochemical conditions. There is a recognizable gradation in brine pool structure between sand-floored (Scotia) and clay-floored (Nulla) salinas. The different hydraulic conductivity of the lake floor sediments changes the ratio of hydraulic conductivity to evaporation rate, which is an important determinant of conditions for brine stability and migration.

4. Salinas generate brine by evaporative concentration and resolution of salt. The structure of the northeast Scotia brine pool suggests that an advective sinking mechanism was operative beneath the late Pleistocene salina.

5. "Fossil" brine has been identified beneath the former, late Pleistocene salina at Nulla. Brine pools migrate in time and space in response to groundwater head fluctuations due to climatic change. These regional groundwater processes have probably operated since the drying of Lake Bungunnia between 400,000 and 700,000 years ago (Stephenson, 1986).

6. Sinking of brines into the large, permeable Pliocene Sand aquifer over a long period of time has contributed to its high regional salinity.

7. The salina known as Nulla Spring Lake is a natural analogue for an effective wastewater evaporation basin. It is likely that long-term containment of saline wastewaters could be achieved here, and in similar clay-floored playas.

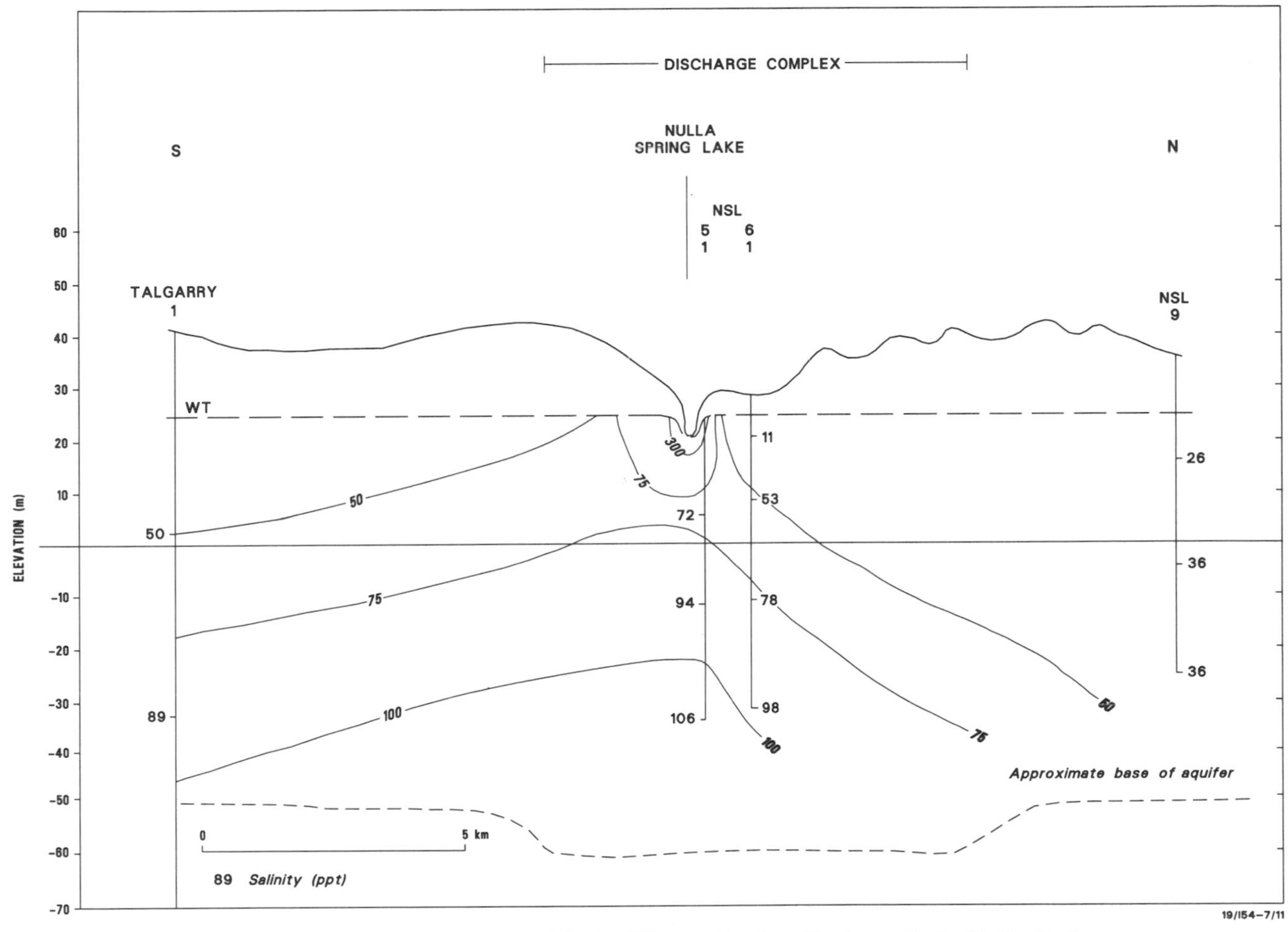

Figure 14. Sinking of fossil brine pool in the Pliocene Sand aquifer beneath the Nulla discharge complex. Bore marked "NSL5" is Arabranch 6 (Fig. 12); bore marked "NSL6" is Arabranch 7 (Fig. 12); bore marked "NSL9" is Arabranch 8 (Fig. 12).

ACKNOWLEDGMENTS

We acknowledge the assistance of the late Campbell M. Brown in developing the geological framework for this project. He was to have contributed to this paper but was tragically killed on a field trip in June 1991. We were assisted in the field by Peter Ryan, Bill Keeley, John Spring, and Mark Glover. We thank Jim Kellett, Jeff Turner, and an anonymous referee for comments on the manuscript. We also thank Robin Wooding, Chris Barnes, Scott Tyler, and Ian White for valuable discussion. Field investigations were partly funded by the Murray-Darling Basin Commission under its Natural Resources Management Strategy. The paper is published by permission of the Executive Director of the Australian Geological Survey Organisation.

REFERENCES CITED

Allison, G. B., and Barnes, C. J., 1985, Estimation of evaporation from the normally 'dry' Lake Frome in South Australia: Journal of Hydrology, v. 78, p. 229–242.

Allison, G. B., and Hughes, M. W., 1983, The use of natural tracers as indicators of soil-water movement in a temperate semi-arid region: Journal of Hydrology, v. 60, p. 157–173.

An, Z., Bowler, J. M., Opdyke, N. D., Macumber, P. G., and Firman, J. B., 1986, Palaeomagnetic stratigraphy of Lake Bungunnia: Plio-Pleistocene precursor of aridity in the Murray Basin, southeastern Australia: Palaeogeography, Palaeoclimatology, Palaeoecology, v. 54, p. 219–240.

Barnes, C. J., Chambers, L. A., Herczeg, A. L., Jacobson, G., Williams, B. G., and Wooding, R. A., 1991, Mixing processes between saline groundwater and evaporation brines in groundwater discharge zones, *in* Proceedings, International Conference on Groundwater in Large Sedimentary Basins, Perth, 1990: Australian Water Resources Council, Conference Series, v. 20, p. 369–378.

Bowler, J. M., 1980, Quaternary chronology and palaeohydrology in the evolution of Mallee landscapes, *in* Storrier, R. R., and Stannard, M. E., eds., Aeolian landscapes in the semi-arid zone of south eastern Australia, Proceedings, Conference, Mildura, 1979: Australian Society of Soil Science, p. 17–36.

Bowler, J. M., 1986, Spatial variability and hydrologic evolution of Australian lake basins: Analogue for Pleistocene hydrologic change and evaporite formation: Palaeogeography, Palaeoclimatology, Palaeoecology, v. 54, p. 21–41.

Bowler, J. M., and Harford, L. B., 1966, Quaternary tectonics and the evolution of the Riverine Plain near Echuca, Victoria: Journal of the Geolog-

ical Society of Australia, v. 13, p. 339–354.
Bowler, J. M., and Teller, J. T., 1986, Quaternary evaporites and hydrological changes, Lake Tyrrell, north-west Victoria: Australian Journal of Earth Sciences, v. 33, p. 43–63.
Brown, C. M., 1989, Structural and stratigraphic framework of groundwater occurrence and surface discharge in the Murray Basin, southeastern Australia: Bureau of Mineral Resources Journal of Australian Geology and Geophysics, v. 11, p. 127–146.
Brown, C. M., and Radke, B. M., 1989, Stratigraphy and sedimentology of mid-Tertiary permeability barriers in the subsurface of the Murray Basin, southeastern Australia: Bureau of Mineral Resources Journal of Australian Geology and Geophysics, v. 11, p. 367–385.
Brown, C. M., and Stephenson, A. E., 1991, Geology of the Murray Basin, southeastern Australia: Australia, Bureau of Mineral Resources, Bulletin 235, 430 p.
Cook, P. G., and Walker, G. R., 1989, Groundwater recharge in south western New South Wales: Centre for Research in Groundwater Processes, Report 9, 31 p.
Davie, R. F., and 7 others, 1989, Chlorine-36 measurements in the Murray Basin: Preliminary results from the Victorian and South Australian Mallee region: Bureau of Mineral Resources Journal of Australian Geology and Geophysics, v. 11, p. 261–272.
Evans, R. S., 1989, Saline water disposal options: Bureau of Mineral Resources Journal of Australian Geology and Geophysics, v. 11, p. 167–186.
Evans, W. R., and Kellett, J. R., 1989, The hydrogeology of the Murray Basin, southeastern Australia: Bureau of Mineral Resources Journal of Australian Geology and Geophysics, v. 11, p. 147–166.
Evans, W. R., Brown, C. M., and Kellett, J. R., 1990, Geology and groundwater. Chapter 5, in Mackay, N., and Eastburn, D., eds., The Murray: Canberra, Murray Darling Basin Commission, Canberra, p. 77–94.
Ferguson, J., Burne, R. V., and Chambers, L. A., 1983, Iron mineralisation of peritidal carbonate sediments by continental groundwaters, Fisherman Bay, South Australia: Sedimentary Geology, v. 34, p. 41–57.
Hills, E. S., 1940, The physiography of Victoria: Melbourne, Whitcombe and Tombs, 292 p.
Jacobson, G., 1988, Hydrology of Lake Amadeus, a groundwater- discharge playa in central Australia: Bureau of Mineral Resources Journal of Australian Geology and Geophysics, v. 10, p. 301–308.
Jacobson, G., Jankowski, J., and Abell, R. S., 1991, Groundwater and surface water interaction at Lake George, New South Wales: Bureau of Mineral Resources Journal of Australian Geology and Geophysics, v. 12, p. 161–190.
Lindsay, J. M., and Barnett, S. R., 1989, Aspects of stratigraphy and structure in relation to the Woolpunda Groundwater Interception Scheme, Murray Basin, South Australia: Bureau of Mineral Resources Journal of Australian Geology and Geophysics, v. 11, p. 219–226.
Macumber, P. G., 1980, The influence of groundwater discharge on the Mallee landscape, in Stornier, R. R., and Stannard, M. E., eds., Aeolian landscapes in the semi-arid zone of southeastern Australia, Proceedings, Australian Society of Soil Science Conference, Mildura, 1979, p. 67–84.
Macumber, P. G., 1983, Interactions between groundwater and surface water systems in northern Victoria [thesis]: University of Melbourne.
Macumber, P. G., 1991, Interaction between ground water and surface systems in northern Victoria: Melbourne, Department of Conservation and Environment, 345 p.
Stephenson, A. E., 1986, Lake Bungunnia—A Plio-Pleistocene megalake in southern Australia: Palaeogeography, Palaeoclimatology, Palaeoecology, v. 57, p. 137–156.
Telfer, A., 1991, Groundwater discharge from a confined aquifer through a zone of high upward leakage, in Proceedings, International Conference on Groundwater in Large Sedimentary basins: Australian Water Resources Council, Conference Series, v. 20, p. 185–193.
Wooding, R. A., 1969, Growth of fingers at an unstable diffusing interface in a porous medium or Hele-Shaw cell: Journal of Fluid Mechanics, v. 39, p. 477–495.
Wooding, R. A., 1989, Convective regime of saline groundwater below a "dry" lake bed: CSIRO Centre for Environmental Mechanics, Technical Report 27, 20 p.

MANUSCRIPT ACCEPTED BY THE SOCIETY JULY 2, 1993

Printed in U.S.A.

Geological Society of America
Special Paper 289
1994

Sedimentation in low-gradient desert margin systems: A comparison of the Late Triassic of northwest Somerset (England) and the late Quaternary of east-central Australia

M. R. Talbot
Geologisk Institutt, Universitetet i Bergen, Allégaten 41, 5007 Bergen, Norway
Kristine Holm
Elf Aquitaine Norge A/S, P.O. Box 168, 4001 Stavanger, Norway
M.A.J. Williams*
Department of Geography and Environmental Science, Monash University, Clayton, Victoria, 3168 Australia

ABSTRACT

The Norian Mercia Mudstone Group of west Somerset (England) is dominated by red dolomitic mudstones and siltstones with thin beds of limestone, sandstone, and nodular or discoidal gypsum or celestite. Primary sedimentary structures are poorly preserved in the clastic units, but evidence of bioturbation and pedogenic features, including rootlet traces, is abundant. In contrast, many limestones contain parallel or ripple lamination. At the time of deposition, peloids seem to have been relatively common in the mudstone and carbonate facies. The sedimentary structures, plus the presence of ooids, *Botryococcus,* and charophyte fragments indicate that the carbonates are of subaqueous origin, but they may also show evidence of gullying, exposure, and desiccation.

Accumulation of the west Somerset Mercia Mudstone Group took place in a low-relief continental basin that had many features in common with the present arid-semiarid interior of Australia. The mudstones and siltstones are of mixed floodplain and playa origin. Some of the sediment was deposited from sheet floods, but material of distant, aeolian origin also made a significant contribution. Rivers entering the basin seem to have carried some of their load as peloidal mud aggregates. The carbonate units represent relatively wet climatic intervals when fresh to brackish lakes occupied shallow depressions on the alluvial/playa plain. Lakes were chiefly maintained by surface runoff, but the presence of spring-related carbonate grains implies that groundwaters may also have been of importance locally. Peloidal micritic carbonate grains in the lacustrine deposits are probably of local aeolian origin and are analogous to the clay pellets found in and around many Australian playas. The evaporites are early diagenetic, in part pedogenic precipitates that formed from saline groundwaters during arid periods. Cyclic alternation of carbonate and mudstone indicate that climatic change may have exerted an important control upon sedimentation during Norian time in southwest England.

*Present address: Mawson Graduate Centre for Environmental Studies, The University of Adelaide, Adelaide, South Australia 5005.

Talbot, M. R., Holm, K., and Williams, M.A.J., 1994, Sedimentation in low-gradient desert margin systems: A comparison of the Late Triassic of northwest Somerset (England) and the late Quaternary of east-central Australia, *in* Rosen, M. R., ed., Paleoclimate and Basin Evolution of Playa Systems: Boulder, Colorado, Geological Society of America Special Paper 289.

INTRODUCTION

During Late Triassic time, sedimentation over large areas of England was characterized by the accumulation of a distinctive set of facies. The dominant lithology is a massive, unfossiliferous, superficially structureless red mud- or siltstone, typically accompanied by evaporites and, around basin margins, by coarser clastic deposits. The fine-grained rocks, the Keuper Marl of earlier authors, can attain considerable thicknesses (>100 m), interrupted only by minor developments of other facies. Such widespread accumulations of apparently featureless mud- and siltstones (see e.g., Ziegler, 1990) have proved particularly difficult to interpret in terms of depositional environment and sediment provenance. A number of environmental settings have been proposed, ranging from deep marine to continental (see reviews in Tucker, 1977; Arthurton, 1980). Although it is now generally accepted that the sediments were not, at least, of deep-water origin there is still no clear consensus as to whether they represent shallow-marine or continental deposits, or perhaps mixtures of the two (Nickless et al., 1976; Tucker, 1977, 1978; Arthurton, 1980; Taylor, 1983; Ziegler, 1990).

The present account will examine the sedimentary characteristics of Late Triassic mudstones, siltstones, and associated rocks in west Somerset (England) and show that illuminating analogies may be made between these deposits and the late Quaternary continental sediments of arid to semiarid central and eastern Australia. We are not the first to compare sediments from the interior of Australia with rocks from the British Triassic. Wills (1970) briefly noted some possible similarities, but numerous detailed, post-1970 studies of Quaternary sedimentation in the Australian interior permit considerably more refined interpretations of the Triassic deposits than were possible at the time Wills did his work. To facilitate the comparison, we will first give a brief overview of the most relevant aspects of the Quaternary of arid and semiarid Australia before proceeding to a more detailed account of the Upper Triassic (Norian) rocks of Somerset.

THE LATE QUATERNARY OF ARID-SEMIARID CENTRAL AND EAST-CENTRAL AUSTRALIA

Quaternary sediments occur widely in the interior of Australia. In eastern areas, where much of the information used in this paper comes from (Fig. 1), these mantle an ancient landscape of subdued relief. Low ridges and cuestas of folded Precambrian and Palaeozoic sedimentary rocks, which rise no more than a few hundred meters above the surrounding plains, form the only major topographic features. Precipitation diminishes from 500 mm annual rainfall at the periphery, to 100 to 150 mm about 1,000 km farther inland.

The Quaternary deposits may conveniently be divided into those that are characteristic of areas marginal to the exposed basement highs, those that are typical of the open plains, and those that occur in both areas.

Marginal deposits

These have been relatively little studied except in the Belarabon area of central western New South Wales (Wasson, 1976; Williams et al., 1991; Fig. 1). Here, valley-fill deposits merge downstream into low-angle alluvial fans that form a discontinuous apron around the flanks of uplands of Paleozoic sedimentary rocks. Both valley and fan deposits display a complex stratigraphy with multiple phases of cut-and-fill and numerous palaeosol horizons, some with significant calcrete development; other sediments, most notably aeolian sands, may separate the alluvial units. Medium to fine sands predominate in the alluvial deposits, coarser, pebbly accumulations being restricted to channel floors. Preservation of primary sedimentary structures is generally very poor, mainly due to extensive bioturbation by insects, which rework the sediments within a few months of deposition. Downslope the fans grade into the extensive plains that dominate the region. Except

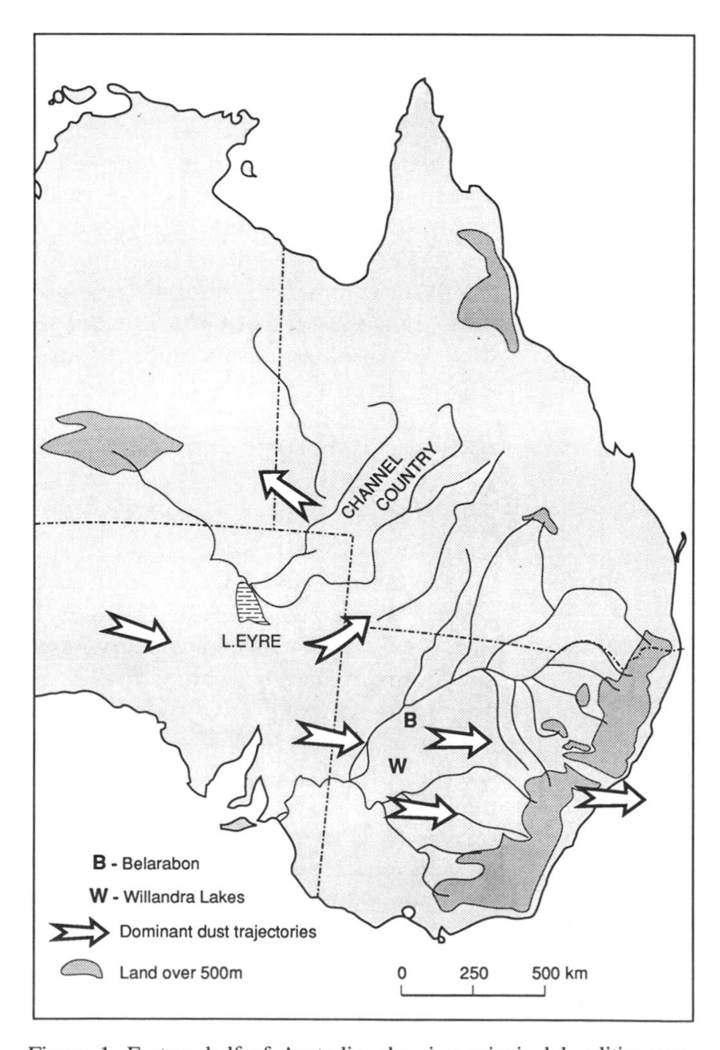

Figure 1. Eastern half of Australia, showing principal localities mentioned in text and dominant trajectories of wind-transported dust (after McTainsh, 1989).

where major perennial streams emerge from the uplands, transition into the sediments of the open plain is typically very rapid, the sands stretching no more than about a kilometer from the valley mouth. Distal portions of the fans carry no clear channels, are composed of very fine sands and silts, and merge into the muds that blanket much of the lowlands. Floods originating in the stream valleys spread out as unconfined sheet flows or follow a maze of shallow, poorly defined, ephemeral channels, the positions of which are principally controlled by vegetation (Fig. 2). The episodic nature of alluvial-fan sedimentation, with evidence for both drier and more humid conditions than exist in the region today, mainly reflects the control exerted by climate change on surface processes in the region (Williams et al., 1991).

Open plains

Despite their superficial monotony, the sediments that underlie the interior lowlands of Australia are complex and have a variety of origins.

Fluvial. Notwithstanding the generally dry climatic regime, alluvial sediments form an important part of the lowland sedimentary cover. The fluvial deposits are to a large degree the products of sheetfloods formed after periods of exceptional rainfall in distant upland areas. Because of the extremely low gradient of the alluvial systems, unconfined flows are able to inundate vast areas once the shallow channel margins have been breached, forming extensive shallow, ephemeral pools of ponded flood water (Woodyer et al., 1979; Rust and Legun, 1983; Nanson et al., 1986, 1988; Fig. 3). The sediment load is overwhelming of mud, much of it transported as sand-sized aggregates derived from the erosion of vertisols (Nanson et al., 1986, 1988; Rust and Nanson, 1989). A considerable proportion of the alluvial mud in the interior of eastern Australia may have originated as mud peloids. In the Channel Country of western Queensland (Fig. 1), Nanson et al. (1988) showed that laterally extensive mud sheets as much as 9 m thick have probably formed from the accumulation of such grains. Discrete aggregates are only discernible in the uppermost 2 to 3 m; below this they gradually disappear, apparently due to merging and disruption during pedoturbation and compaction. The end product is a structureless mud with a clotted fabric, within which float silt- and fine sand-sized mineral grains (see, e.g., Nanson et al., 1988, Fig. 14). Sand accumulations comprise only a minor part of the total sediment, occurring mainly as isolated channel fills (Rust, 1981; Rust and Nanson, 1989).

Lacustrine. Lake basins are widely distributed across the eastern interior of Australia. Many of them are at present ephemeral or playa systems, but they have clearly held significant perennial bodies of water during periods of the late Pleistocene and Holocene. The origin of the basins varies. Some are deflation hollows and others have formed on areas of abandoned flood plain; the largest are of tectonic origin. Lake Eyre, for example, which is the terminal basin for much of the south- and westward-directed drainage from the uplands of northeastern Australia, is an intracratonic sag (Dulhunty, 1982; Fig. 1). Virtually all the lakes in the region are shallow, with a tendency to be transformed into saline pans or playas during dry periods. The dominant ions are sodium, calcium,

Figure 2. Shallow, ephemeral flood channels on plains west of Belarabon, New South Wales.

Figure 3. Desiccated ephemeral pond muds veneered by muds that settled out from ponded flood waters. Emu footprints. Belarabon, New South Wales.

chloride, sulphate, and bicarbonate. Many aspects of the history, hydrology, sedimentology, and biota of these lakes have been reviewed by Bowler (1981, 1986), De Deckker (1983, 1988), Torgersen et al. (1986), Teller and Last (1990), and Williams et al. (1991).

Phyllosilicate clays and other detrital silicate minerals dominate the fill of many basins (Bowler, 1973; Bowler and Teller, 1986). These, together with the authigenic components described below, can reach considerable thicknesses. For example, Williams et al. (1991) have demonstrated, by augering, the presence of at least 14 m of brown clay beneath the Bunda playa lake, Willandra, in western New South Wales (Fig. 1). They infer from seismic evidence that the total thickness is in excess of 50 m. Such sequences are unlikely to reflect continuous lacustrine sediment accumulation, however. Detailed investigations of cores usually reveals the presence of numerous hiati, typically marked by soil horizons, which indicate desiccation of the basin and removal of sediment by deflation (Bowler, 1986; Bowler and Teller, 1986; Fig. 4).

Figure 4. Early Holocene brown, lacustrine silts covering soil that developed during exposure of earlier lacustrine deposits. The latter are extensively disrupted by desiccation and bioturbation. B, burrows; R, root traces; M, layer of molluscs at base of Holocene silts. Boolaboolka playa, western New South Wales (photo by D. A. Adamson).

Authigenic sedimentary components provide a significant proportion of the fill in many of the basins. The water chemistry favors the widespread formation of carbonate and gypsum deposits; calcite and/or dolomite seems to be present in virtually all the lakes that have been studied to date and gypsum is common in many. Ephemeral halite also occurs, but significant accumulations of bedded halite are of limited importance (Bowler, 1986). Carbonate sediments may be of biological or abiological origin. Shell beds formed from the accumulation of molluscan or ostracode shells are widespread (Fig. 4); carbonate-impregnated charophyte debris may also be a significant component of the sediments in some lakes (De Deckker, 1988). In addition, carbonate muds of probable chemical origin are common and calcite ooids are also recorded in some basins, the latter being particularly characteristic of periods of high water (De Deckker, 1988). Microcrystalline dolomite is a ubiquitous component of the muddy sediments of these basins (De Deckker and Last, 1989; Last, 1990; Teller and Last, 1990).

A characteristic feature of the margins to many of the lakes are clay mantles and lunettes—arcuate dune ridges—built mainly from the accumulation of clay or clay-carbonate peloids (Bowler, 1973, 1976, 1981; Bowler and Teller, 1986; De Deckker, 1988; Williams et al., 1991; Fig. 5). Clay pellets are typical of ephemeral, low water conditions in the basin, when they form during seasonal drying due to the combined effects of surface efflorescence and deflation of the playa sediments (Price, 1963; Bowler, 1973). Once dry, the peloids are relatively resistant (Fig. 5) and may be reworked back into the lake during subsequent periods of dissection of the lunettes (Bowler and Teller, 1986).

Gypsum occurs widely, both as displacive crystals scattered within other lake sediments or as beds of discrete crystals. The former are of early diagenetic origin, while the latter

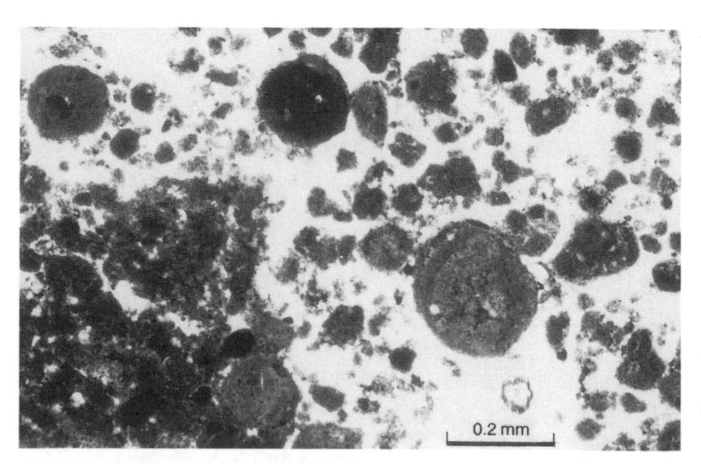

Figure 5. Photomicrograph of Holocene aeolian clay mantle, showing fine, peloidal structure of the clay. Individual peloids range from 0.1 to 0.5 mm in diameter. The clay mantle overlies vertisolic clays of a late Pleistocene flood plain.

formed as accumulations of primary precipitates from the open lake waters or *via* reworking of earlier gypsum-bearing units (Warren, 1982; Bowler and Teller, 1986).

Aeolian. In addition to the clay peloid lunettes, the most tangible evidence of the importance of aeolian processes in the interior of Australia is provided by the occurrence of major dune systems—it is estimated that 40% of the Australian desert surface is covered by wind-blown sands. Some of these are active at the present-day, but dune formation was much more widespread during the late Pleistocene and Holocene arid phases (Brookfield, 1970; Wasson, 1976, 1983, 1986; McTainsh, 1989; Williams et al., 1991). Dunes occur in association with all the sediments described above; in many areas their formation is clearly related to the existence of a plentiful sand supply provided by sand-transporting alluvial systems (e.g., McTainsh, 1989; Williams et al., 1991).

Other, less obvious aeolian sediments are possibly of equal volumetric importance to the dune sands. Very large areas marginal to the arid core of Australia are blanketed by a silty clay, much of which originated as loess (*parna* of earlier authors: Butler, 1956; Dare-Edwards, 1984; McTainsh, 1989). Accumulation rates can be extremely high immediately downwind of the dust source areas (Fig. 1)—McTainsh (1989, p. 250) indicates figures in excess of 10 cm/yr—and were in all probability exceptionally high during the drier, windier periods of the late Pleistocene. The loessic blanket consists of a mixture of clay and silt- to sand-sized mineral grains. Like the clay dunes that surround many lake basins, the clay fraction is thought to have originated largely as clay peloids, entrained as silt-sized particles from the dried-out surfaces of playas and flood plains (Dare-Edwards, 1984).

Figure 6 summarizes the principal sedimentary environments of continental central-eastern Australia during the Quaternary.

THE UPPER TRIASSIC OF WEST SOMERSET

The rocks to be discussed here belong to the Mercia Mudstone Group (formerly Keuper Marl) and are of probable Norian age (Warrington et al., 1980). Although present over extensive areas of west and central Somerset (Fig. 7), they are generally poorly exposed and can be studied satisfactorily only in cliff and beach exposures at various points along the Bristol Channel coast (Fig. 7), where the succession has a maximum exposed thickness of about 67 m (Whittaker and Green, 1983). These coastal sections have been the main source of data for the present account. In addition, material from similar stratigraphic levels has been examined from the core of the British Geological Survey borehole at Brent Knoll (Whittaker and Green, 1983; Fig. 7).

The west Somerset Norian accumulated along the southern margin of a basin oriented roughly east-west, the Bristol Channel Basin, an offshore continuation of the Central Somerset Basin (Kamerling, 1979; Whittaker and Green, 1983; Fig. 8). Although basins in southern Britain were active, fault-

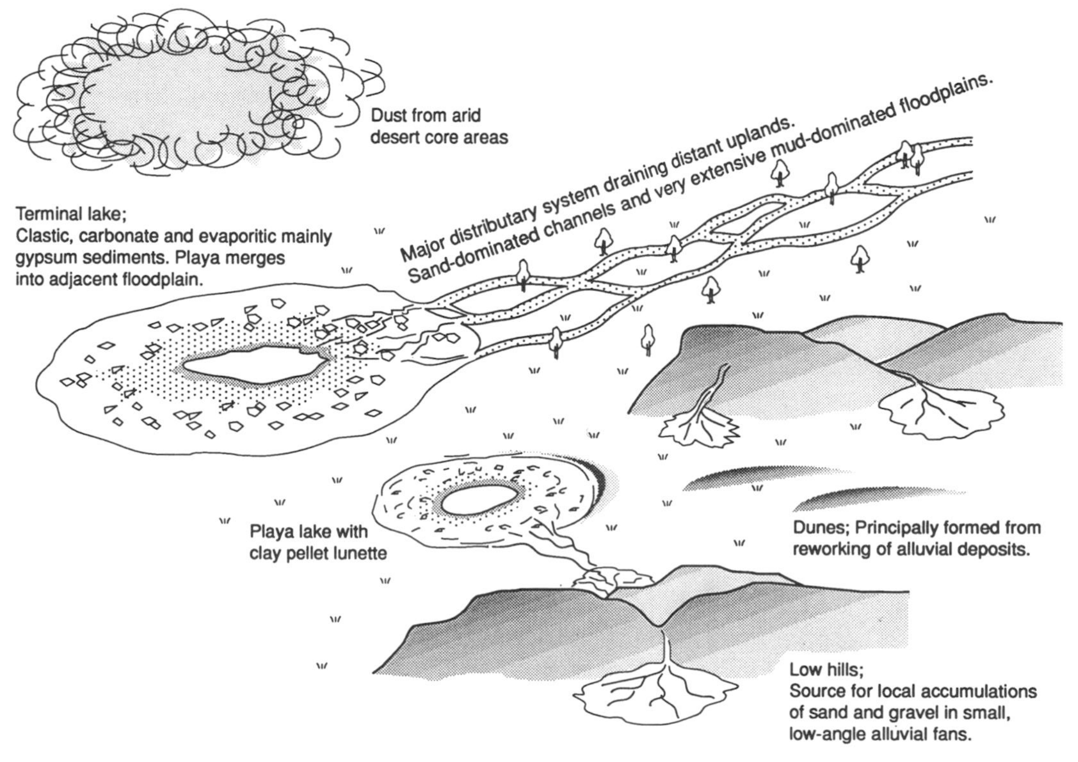

Figure 6. Principal sedimentary environments of the desert-margin areas of interior eastern Australia.

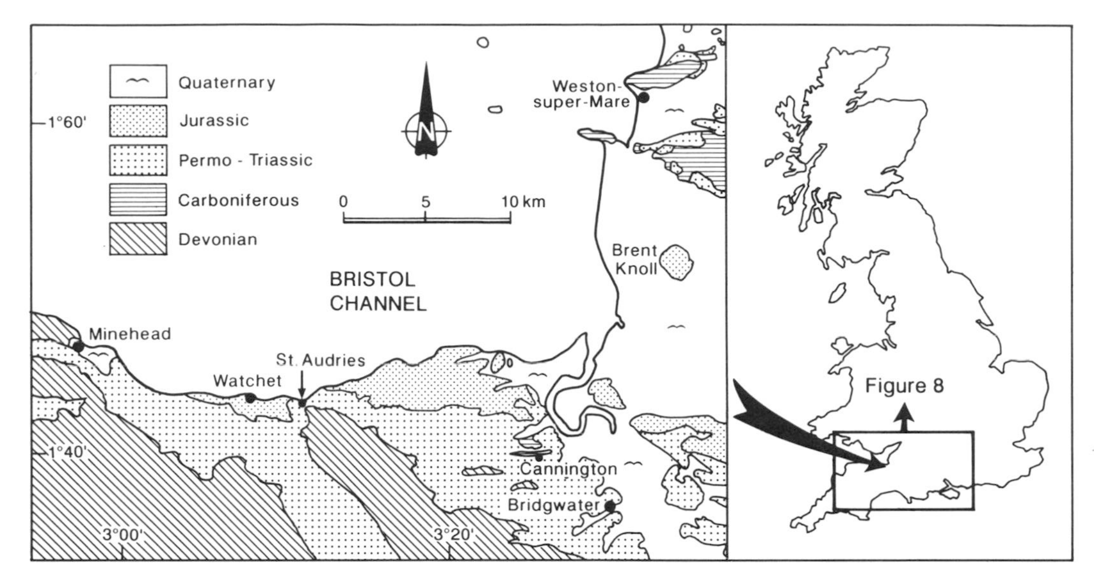

Figure 7. Simplified geological map of west Somerset. Principal sections of the Mercia Mudstone Group in this region are along the coast in the Watchet and St. Audries area.

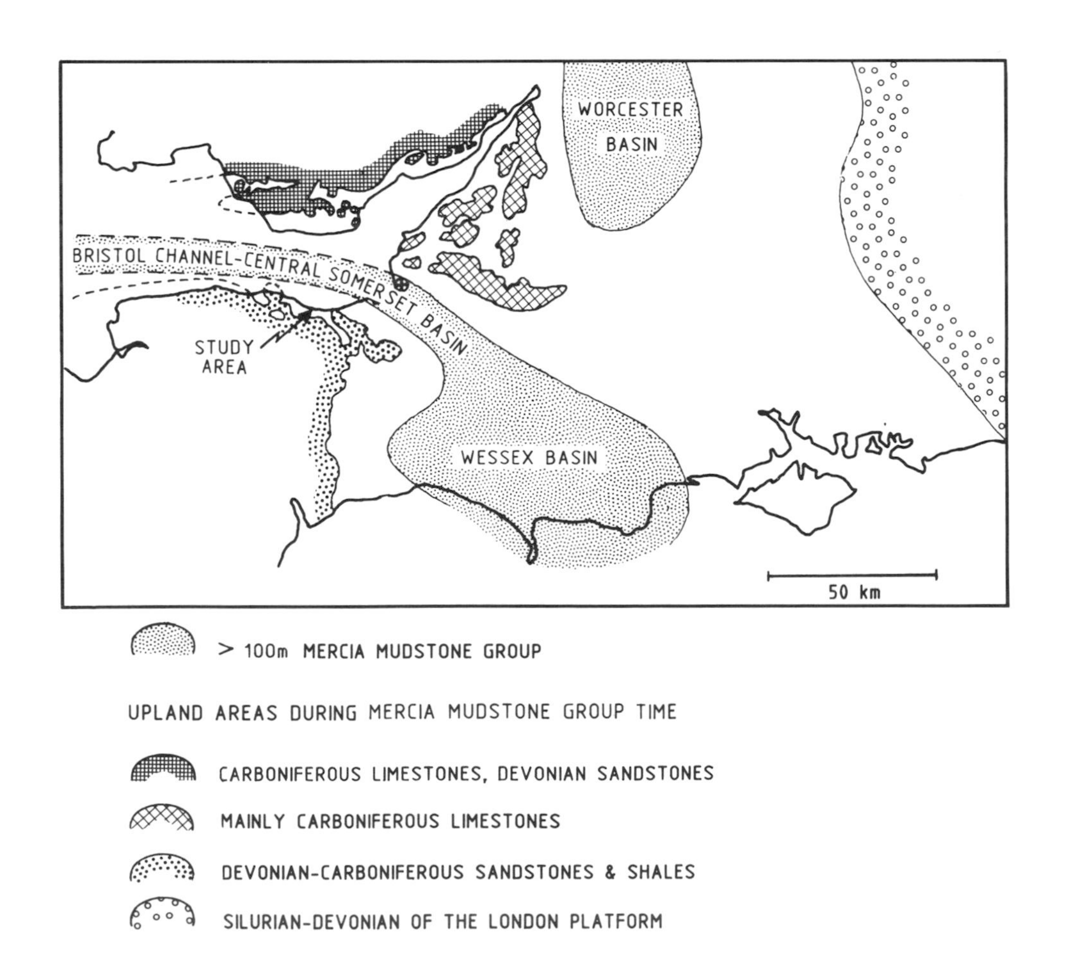

Figure 8. Late Triassic palaeogeography/palaeogeology of southwest Britain, showing position of principal Mercia Mudstone Group basins.

bounded structures of rift-graben type during Early to mid-Triassic time (Kamerling, 1979; Holloway, 1985; Chadwick, 1986; Ziegler, 1990), there is little exposed evidence of Norian faulting around the margins of the Bristol Channel Basin. Here sedimentary contacts are the rule, the Mercia Mudstone Group (MMG) or its marginal equivalents overlapping Permo-Triassic sandstones and conglomerates to rest unconformably upon folded upper Palaeozoic rocks. The MMG of west Somerset was evidently deposited during a period of relative tectonic quiescence, an observation that apparently holds true for much of southern Britain, since overlapping stratigraphic relationships predominate (Holloway, 1985) and Chadwick (1986) found evidence for reduced fault activity in the region at that time.

Palaeozoic rocks formed minor uplands flanking the basin and were the source of a coarser, marginal facies (the "Dolomitic Conglomerate": Green and Welch, 1965; Curtis, 1982; Whittaker and Green, 1983; Berge, 1985). As previous workers have noted, distribution of these coarser rocks is very localized. Generally, they are confined to a narrow marginal zone that typically stretches no more than a few kilometers into the basin from the point where the distributary system emerged from the uplands. Thus, mudstones and siltstones dominate the basin fill, together with evaporites, which reach considerable thicknesses in unexposed central areas of the basin (Kamerling, 1979; Whittaker, 1980). In addition, the sequences to be discussed here contain a number of distinctive carbonate units that provide valuable information upon the probable environments of deposition of the associated fine clastic sediments.

Lithofacies

Representative sections from the MMG at various points along the west Somerset coast are shown in Figure 9. In this account we will discuss four major lithofacies: mudstones and siltstones, sandstones, carbonates, and evaporites. The logs shown in Figure 9 are based upon field investigations supported by thin-section petrographic studies using conventional transmitted light and incident ultra-violet (UV) light microscopy (Talbot, 1984; Dravis and Yurewicz, 1985).

1. Mudstones and siltstones. Volumetrically these are by far the most abundant lithologies in the MMG (Figs. 9, 10). At outcrop, they appear typically as red and less commonly as green unfossiliferous deposits with blocky structure and are generally poor in primary or macroscopic biogenic sedimentary structures. Horizontal bedding is in fact widespread but is commonly rather vague and discernible only after close inspection. The original sediments evidently possessed centimeter- to decimeter-scale horizontal lamination, due mainly to small variations in grain size, but post-depositional processes (see below) have strongly disrupted the primary depositional textures of these rocks (Figs. 10d, 11). Small-scale lamination is normally visible only where these fine clastics are interbedded with the other lithologies described below or, less commonly, over limited stratigraphic intervals where halite casts occur. Here the sediments may display a very fine, uniform lamination. Elsewhere, blocky structures predominate, the rock tending to break up into equidimensional, centimeter-scale angular to subangular blocks. Red clay films coat some block surfaces. Locally, large-scale early post-depositional disruption of the sediments is apparent in the form of desiccation polygons, which can be up to 1.5 m across. Red mudstone typically abruptly overlies such horizons and may penetrate considerable distances down the cracks.

Texturally, the sediments are mudstones or fine to medium siltstones. The mudstones invariably contain some silt-size grains, typically a mixture of angular to subangular quartz with subordinate feldspar and a variety of carbonate grain types. The most characteristic carbonate grains are 20 to 80 μm dolomite rhombs, which occur scattered amongst the other grain types or concentrated in specific zones (Fig. 11). Truncated compositional zoning, revealed by ultraviolet (UV) and cathodoluminescent (CL) microscopy, indicates that some dolomite grains are detrital, but diagenetic dolomites are almost certainly present, too (e.g., Tucker, 1978; Taylor, 1983). In west Somerset other carbonate grain types also occur. Most common are silt-to fine sand-size clasts derived from a variety of limestones, some fossiliferous (Figs. 11a, 12a), and fragments of dense micrite, some having a concretionary fabric (Fig. 13b). Clusters of small calcite crystals occur as radial growths or sheetlike aggregates with one planar surface and one surface formed of pyramidal or spherulitic calcite (Fig. 13a).

A much less common occurrence of coarser grain within the mudstones is as laminae, some just a single grain thick, or isolated, outsize clasts of very well rounded, medium sand-size grains of quartz, feldspar, or limestone (Fig. 14c). Scattered, tabular gypsum crystals are associated with some of these sand-bearing horizons and are evidently of early diagenetic origin as the crystals have subsequently been deformed by compaction (Fig. 14c).

The coarser clastic grains are set in a matrix of, and commonly impregnated by, red-colored, very fine silt to clay. Illite, chlorite, corrensite, smectite, and sepiolite dominate the clay mineral assemblage (Dumbleton and West, 1966; Jeans, 1978; Taylor, 1983). The various grain types are not homogeneously distributed and it is clear from thin-section examination that a variety of structures and fabrics are present which are not visible in hand specimen. At least some of the mud was originally present in peloid form (Fig. 11c) and many of the mudstones have a vague, clotted texture, suggesting that peloids may formerly have been much more abundant in these rocks (Fig. 11). Clay-grade material also occurs as coatings to larger grains, as the filling to voids that may branch and die out downwards (Fig. 14b) and defines meniscuslike or concave-up microlaminae, some of which clearly form the fill to cavities of probable biogenic origin (Fig. 14a).

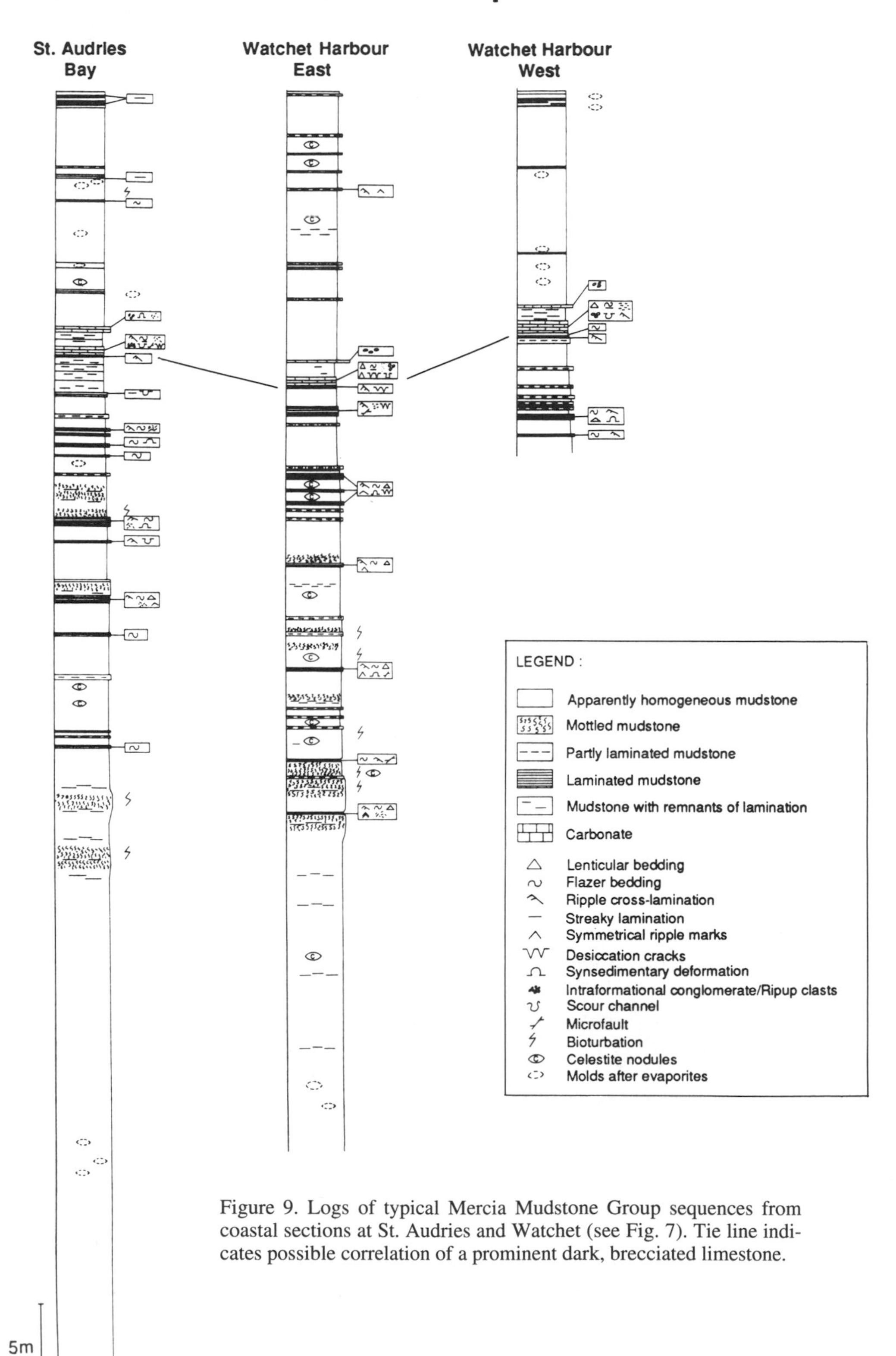

Figure 9. Logs of typical Mercia Mudstone Group sequences from coastal sections at St. Audries and Watchet (see Fig. 7). Tie line indicates possible correlation of a prominent dark, brecciated limestone.

Figure 10. a, Outcrop of upper part of Mercia Mudstone Group, 200 m east of Watchet Harbour, showing interbedded carbonate (light) and red mudstone units. Stratigraphic thickness shown is about 50 m. b, Top of a carbonate unit with several generations of oscillation ripples (R) and traces of desiccation features (D). c, Vertical section through upper part of a laminated carbonate unit. Synsedimentary brecciation is apparent at a number of horizons and the upper part of the unit shows evidence of desiccation and gullying. Figure 11d comes from one of the brecciated units. d, Thin, bioturbated sand (just above top of ruler) resting upon red mudstone. Unusually large burrows (B) are present in the left-hand side of the photograph and a large desiccation crack (D) at right center. Most of the wispy, irregular light-colored structures are rootlet traces.

Interpretation. At present there is no single, generally accepted explanation for the origin of the red mudstone/siltstone facies of the Mercia Mudstone Group (see Introduction). Interpretation has been hampered by the almost total absence of fossils and the scarcity of primary sedimentary structures. Current depositional models for the mudstones fall into two groups: (a) essentially subaqueous accumulation in lacustrine basins (Warrington, 1974; Tucker, 1977, 1978); or (b) accumulation upon generally subaerial, playalike flats, with much of the fine sediment being supplied by aeolian processes (Lomas, 1907; Taylor et al., 1963; Wills, 1976; Curtis, 1982) or *via* low-angle alluvial fan systems (Wright et al., 1988; Wright, 1992). With particular reference to the Bristol Channel basin, Tucker (1977, 1978) has noted that along the northern margin the mudstones are laterally transitional into coarser, littoral facies and suggests that the former represent the offshore, deeper water deposits of a large lake. The association with evaporites is explained in terms of climatic variations, the mudstones accumulating under relatively more humid conditions with high lake levels, while evaporitic sections represent periods of rather arid climate, with low, saline waters in the basin (Tucker, 1978). In view of the preponderance of mudstone, it would seem that high lake levels must have dominated the basin, in which case it must be noted that the massive red mudstones constitute a highly atypical open lacustrine sedimentary facies. Quaternary climatic variations have had a major impact upon existing mid- to low-latitude lakes, causing dramatic changes in lake level (see, e.g., Street-Perrott and Roberts, 1983; Talbot, 1988). Basinal sediments, some with associated evaporites, exposed by these fluctuations nevertheless retain clear indications of their origins. Sedimentary features such as fine lamination, relatively abundant organic matter, evidence of frequent wave action, primary carbonate laminae, and traces of bioturbation by aquatic benthos are all common in the Quaternary examples (e.g., Begin et al., 1974; Valeton, 1978; Hardie et al., 1978; Tiercelin et al., 1981; Petit-Maire, 1982, 1986; Ritchie et al., 1985), but rare or absent from the mudstone facies of the MMG of the Bristol Channel basin. Neither does this facies dominate time-equivalent lacustrine sequences in the early Mesozoic rift basins of eastern North America. Although present, massive red mudstones form only part of a complex suite of facies comparable to that found in Quaternary lacustrine basins (Van Houten, 1962, 1964; Olsen, 1984; Demicco and Gierlowski-Kordesch, 1986; Gore, 1988; Smoot and Olsen, 1988; Smoot,

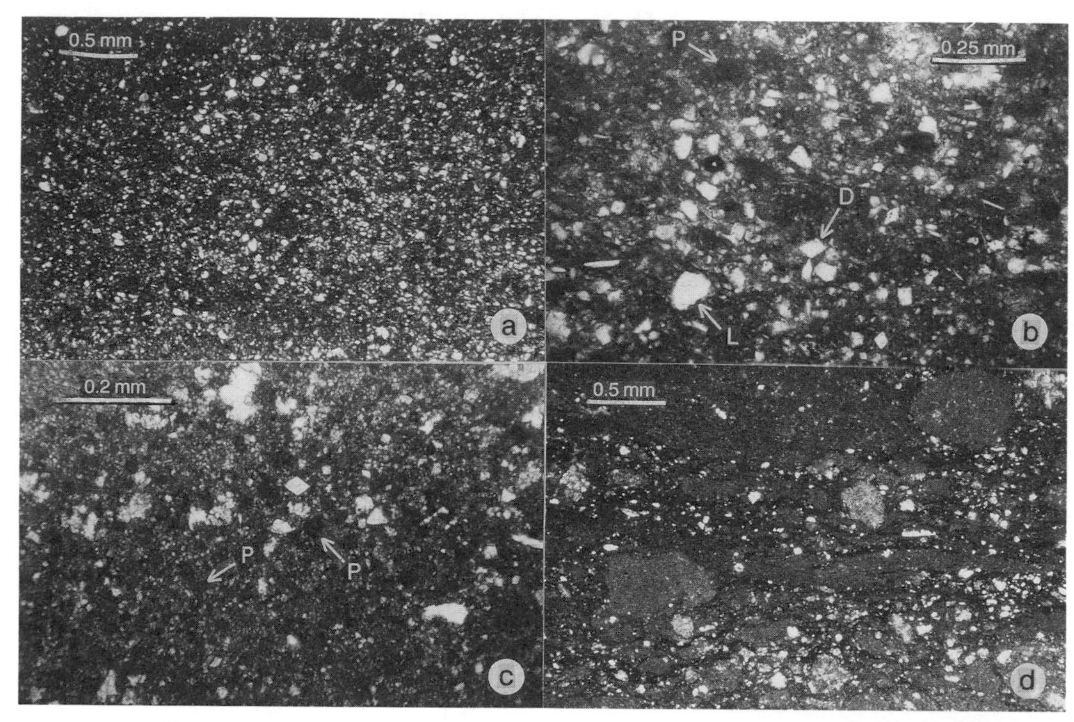

Figure 11. Characteristic features of some typical Mercia Mudstone Group Facies. a, Siltstone showing clotted fabric. b, Muddy siltstone. Clasts are mainly quartz, but isolated, probably clastic dolomite (D), limestone clasts (L), and hints of former mudstone peloids (P) are also present. c, Mudstone with abundant mud peloids (P). Note relatively good sorting of peloids and similarity with size of clastic quartz and dolomite grains. d, Micritic carbonate with intense brecciation fabric (sample is from bed shown in Fig. 10d).

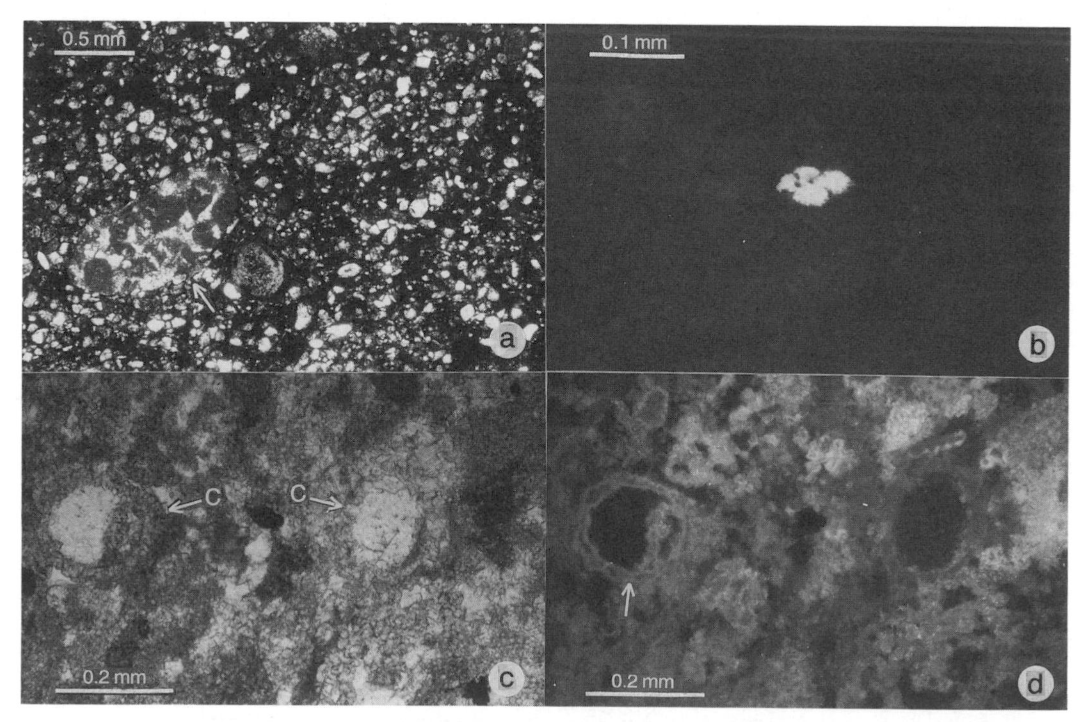

Figure 12. Organic remains in Mercia Mudstone Group rocks. a, Fossiliferous Carboniferous Limestone clast (arrowed) in siltstone. b, *Botryococcus* colony in micritic carbonate. Ultraviolet incident light. c, Probable charophyte gyrogonite (C) in fenestral microsparite. Plane-polarized light. d, same as photo C under ultraviolet light showing characteristic double wall (arrowed) of gyrogonite.

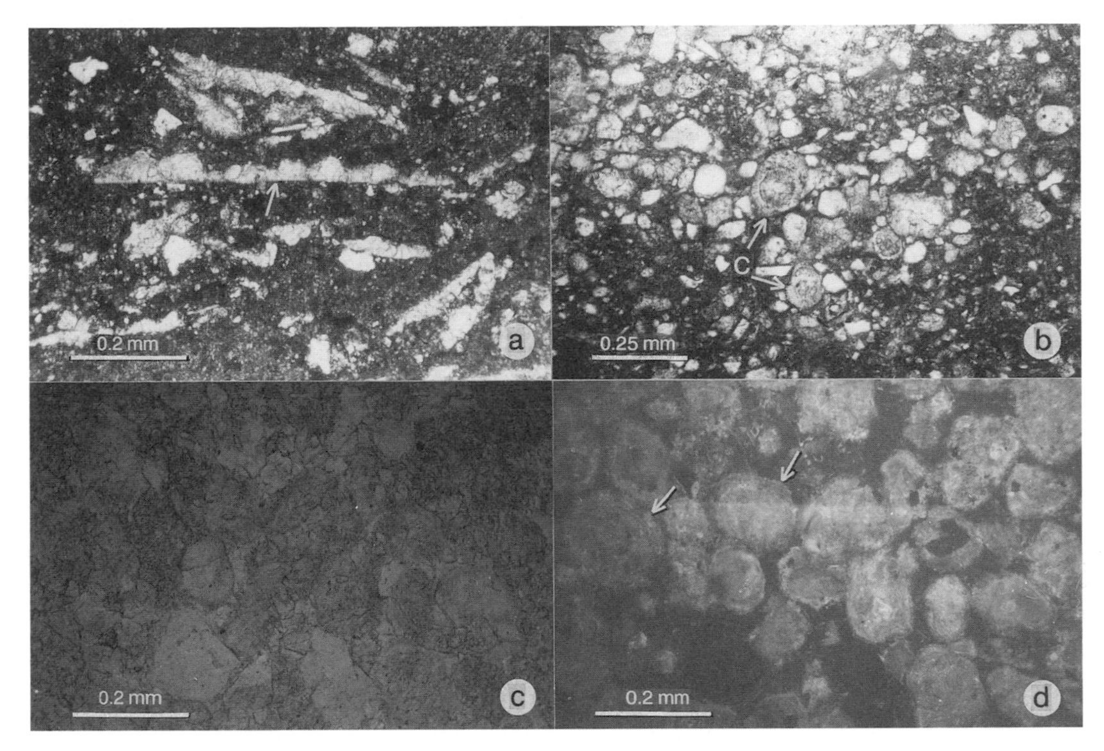

Figure 13. Carbonate grains. a, Calcite crust fragments. Note asymmetry of several fragments with one flat side (arrowed) and the other showing irregular growth of spherulitic calcite. b, Grains of spherulitic and concretionary calcite (C), possibly reworked from calcareous soil profiles. c, Dolomitic microspar mosaic, which under ultraviolet (d) is seen to have been an ooid grainstone fabric (arrows indicate well-developed ooid cortices).

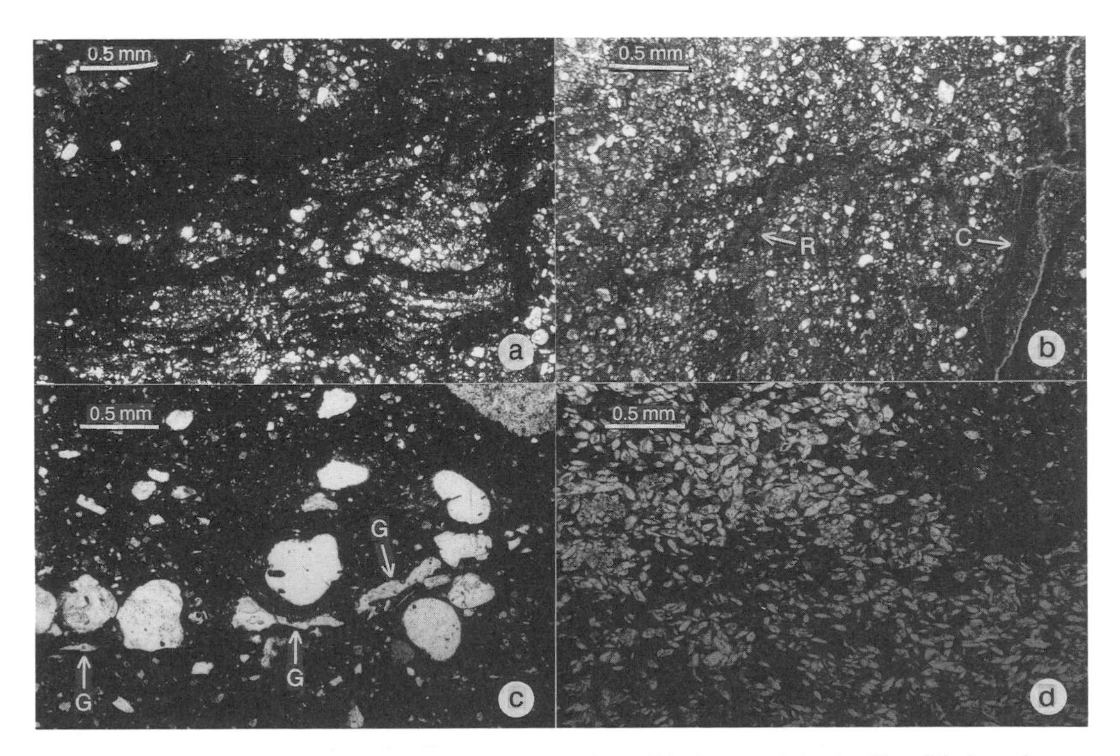

Figure 14. a, Meniscus and back-filled burrows of possible insect origin. b, Clay-filled rootlet traces (R) and clay-lined crack (C) in siltstone. c, Single grain layer of well-rounded outsize quartz grains in mudstone. Discoidal gypsum crystals (G), some deformed by compaction, are associated with such layers. d, Discoidal gypsum crystals in red mudstone.

1991). It could perhaps be argued that post-depositional processes have destroyed characteristic open-lake features in the MMG, but that essentially all should have been so effectively removed seems highly improbable. Pedogenic processes have indeed changed some of the exposed Quaternary sediments referred to above; Hardie et al. (1978, p. 21), for example, note that in the exposed sediments from a former high stand of the Great Salt Lake (Utah) "the uppermost 20-30 cm, which is a highly churned soil-like zone, shows evidence of recent exposure to mudcracking, salt-crust deformation, oxidation, insect burrowing and, in places, rooting by halophytes." Finely laminated open lake sediments are well-preserved beneath this surface zone, however. Similar transitions are apparent in many of the lacustrine cycles of the early Mesozoic Newark Supergroup, deposits typical of high lake levels (commonly dark, organic-rich laminated mudstones) are nevertheless well preserved beneath the massive, churned mudstone facies (Van Houten, 1964; Olsen, 1986; Smoot and Olsen, 1988). An offshore lacustrine origin for the red siltstones and mudstones of the MMG therefore seems unlikely.

In the alternative, terrestrial model, the fine-grained facies are thought to have accumulated on moist or dry surfaces. Formation of "ploughed" or "puffy ground" by the precipitation of salts from groundwaters at or near the surface, has been advanced as an explanation for the absence of lamination (Hardie et al., 1978; Arthurton, 1980; Kendall, 1984). While also favoring a predominantly terrestrial origin for the red mudstone facies, we feel that the evidence does not support a dominance of saline groundwaters, at least along the basin margin. Gypsum-bearing horizons make up only a minor part of the succession (Fig. 9; see below for a more detailed discussion) and although lamination is indeed poor or absent from these, it is equally rare in mudstones lacking all trace of evaporites. Where halite casts occur, lamination may be disrupted, but as we have noted above, these horizons can also be notable for the preservation of fine lamination, so disruption by salt crystal growth seems an unlikely explanation for the common absence of primary sedimentary structures. Many of the textures and fabrics observed in the Somerset MMG are typically pedogenic. In particular, the common blocky structure (peds) with clay coatings (argillans) and on a microscopic scale, the widespread development of clay coatings (cutans) on coarser grains, of clay-filled voids, and anastomosing and branching clay films (Fig. 14b), which have probably replaced root hairs, are all characteristic features of soils (Brewer, 1976; Birkeland, 1984; Retallack, 1983, 1988; Gustavson, 1991). The small, meniscus-filled burrows, commonly associated with these fabrics (Fig. 14a) are typical for terrestrial deposits and have probably been produced by insects (J. Pollard, personal communication, 1987). Additional evidence of soil-forming processes, in particular vertisol features, has been recognized in Somerset and adjacent areas (Wright et al., 1988; Wright, 1992). Thus, far from being deposited on barren, saline flats, we suggest that much of the mudstone/siltstone facies accumulated on surfaces that were at least periodically able to support vegetation and allow the establishment of a soil microfauna. The churning effects of plants and animals and seasonal wetting-and-drying cycles are probably responsible for the absence of primary structures and the development of the characteristic blocky structure and clay coated peds. Some of the plants may well have been halophytes—the coexistence of rooted halophytes, burrowing insects, and surface salt efflorescence is by no means uncommon at the margins of playas (De Deckker, 1988; Rosen, 1991; MRT, personal observation, Great Salt Lake)—but the complete absence of evidence for evaporite minerals through significant thicknesses of the red mudstones leads us to conclude that saline-soil conditions were by no means a constant feature of the mudstone environment. Preservation of fine lamination in association with some of the halite cast-rich horizons, on the other hand, reflects periods of hypersaline conditions, probably in ephemeral ponds, which hindered sediment colonization and thus prevented bioturbation. Evidence for dry surface conditions is provided by the laminae of outsize sand grains (Fig. 14c), which were emplaced by aeolian creep and saltation across thinly vegetated, desiccated soil surfaces. Elevated soil water salinities may well have developed during these periods; the gypsum laths that are associated with the wind-blown sand laminae strongly resemble pedogenic gypsum that Magee (1991) has recorded from some southeastern Australia Quaternary lacustrine sequences.

Wright (1992) suggests that mud accumulation occurred on low-angle alluvial fans. This was very probably the case in basin-margin areas, but the best modern analogue for the bulk of the Triassic mudstones is perhaps provided by muddy, ephemeral flood plains of the sort that cover vast areas of the interior of eastern Australia. There are many striking similarities between the sediments of this region and those of the MMG mudstone/siltstone facies, notably the dominance of fine-grained clastics and pervasive pedogenic textures and microfabrics (cf. Fig. 11b with Fig. 14d in Nanson et al., 1988). We have previously noted the widespread occurrence of mud peloids in the Australian interior. To what degree the MMG sediments were transported as mud aggregates is now impossible to say; traces of peloids are discernible at several horizons (Fig. 11) and the widespread clotted fabric suggests peloids may have been common; the low preservation potential of mud peloids has probably effectively eliminated conclusive evidence for their former widespread occurrence. Nevertheless, we feel that sediment deposition occurred on wide, low gradient, partially vegetated plains which were periodically inundated by floods emanating from the adjacent Palaeozoic uplands. The most important sedimentary processes were probably accumulation of mud aggregates, transported as bed-load (Nanson et al., 1988; Rust and Nanson, 1989) and settling of suspended sediment from ponded flood waters. Periods of standing water or a shallow groundwater table may have favored the formation of dolomite, magnesium being supplied by dissolution of clastic grains and by release from Mg-bearing clays. Reworked dolomite rhombs would have come from the erosion of earlier-deposited flood plain and lacustrine deposits (see below). A further significant source of sediment was probably wind-blown dust (see below). Low-angle fans of the type envisaged by Wright (1992) merged into proximal parts of these plains (cf. Williams et al., 1991) and it is probable that during periods of their existence, parts of the flood plains passed laterally into playas, just as the Channel Country merges into the northern playa of Lake Eyre (Dulhunty, 1982). An analogous situation occurs along the northeast margin of Lake Urmia (Iran), where the low-gradient (3/5,000) Talkeh-rud River enters the basin, depositing vast flats of red mud that are gradational with the lake's marginal playa. Significantly, similar red muds beneath the present carbonate sediments of Lake Urmia are dolomitic (Kelts and Shahrabi, 1986). Within the MMG, rocks of playa origin are most likely to be represented by the red mudstones and siltstones interbedded with lacustrine carbonates and some of the evaporite-bearing horizons (see below).

Other carbonate clasts present in the mudstones provide additional information on sediment sources and transport directions. Many of the limestone clasts are of typical Carboniferous Limestone lithologies. The closest outcrop of these rocks is at Cannington (Fig. 7), where they occur as a small inlier surrounded by Mesozoic rocks. Extensive outcrops were present along the northern and eastern flanks of the Bristol Channel Basin (Fig. 8), but are not known west of the study area. These clasts therefore suggest derivation of at least some of the MMG from the east or north, indicating the possible existence of an axial drainage system from that direction. The fragments of concretionary limestone (Fig. 13b) resemble pedogenic carbonates (cf. Hay and Wiggins, 1980). Calcareous soils occur in the South Wales MMG (Tucker, 1977; Leslie et al., 1992) and are likely to have been widely developed upon Carboniferous limestone uplands surrounding the basin (e.g., Tucker, 1977). Aggregates of calcite crystals with a crustose habit (Fig. 13a) probably formed as precipitates at the surface of shallow pools. Such crusts are especially characteristic of calcite-supersaturated waters at emergent springs (Chafetz et al., 1991). The fragments seen in this study must have been reworked from spring orifices. Spring tufas and related deposits have been identified from marginal equivalents of the MMG in South Wales (Leslie et al., 1992) and were probably also formed along the south side of the basin. Like today, the Palaeozoic rocks in this region, especially the Carboniferous Limestone, would have been an important source of groundwater (Simms and Ruffell, 1990; Leslie et al., 1992). Springs may have made a significant contribution to the hydrology of the Bristol Channel Basin during Late Triassic time.

*2. **Sandstones.*** Two sandstone types have been recognized. One occurs as thin, isolated beds within the mudstones, the other as more substantial units representing one or more significant periods of sand accumulation.

(i). Isolated sandstone beds. These are units of fine to medium sand up to 25 cm thick, bounded above and below by typical red mudstone (Fig. 9). The basal contact may be sharp, but more commonly there are gradational top and bottom contacts, due to post-depositional mixing. Irregular parallel or small-scale current ripple lamination forms the commonest primary sedimentary structures. In general, however, such structures tend to be poorly preserved, due to intense bioturbation that has destroyed the lamination and mixed sand with the adjacent mudstones (Fig. 10d). The burrows are generally undiagnostic, irregular, vertical to horizontal curved tubes, 2 to 5 mm in diameter, with rare traces of a meniscate back-fill structure. Burrows of this sort are typical of the *Scoyenia* ichnofacies, a characteristic terrestrial trace fossil assemblage (Frey et al., 1984; Blodgett, 1988; Olsen, 1988).

Interpretation. Isolated, thin sheet sandstones with (where preserved) a sharp, locally erosional base are probably the product of rare, high-intensity flood events that caused exceptional discharge from the marginal fans and transported coarse clastic sediments much farther into the basin than was normal. Subsequent colonization of the resulting sand sheets by plants and burrowing organisms disrupted internal sedimentary structures and blurred the originally sharp basal contact.

(ii). Thick sandstones. More substantial sand accumulations within the west Somerset MMG have been recorded at only one locality, west of Watchet. Here an approximate 2 m thick sandstone unit has a sharp, but not obviously erosional contact with the underlying mudstone, above which comes a zone of interlaminated, well-sorted sand- and siltstones with ripple lenticles of medium to coarse sand, followed by a deformed interval. Within the main body of the sandstone, the lamination varies from well preserved to indistinct and is dominated by parallel to gently undulating laminae with scattered low-angle climbing ripple forms. The top of the unit is churned, has a patchily reddened appearance and scattered calcareous nodules. There are indications that bioturbation is at least in part responsible for the disruption, but well-preserved burrow structures are rare. Quartz and minor feldspar are the dominant minerals; in contrast to the mudstones, clastic carbonate grains are uncommon. Many of the larger sand grains are very well rounded. Much of lower part of the unit is cemented by poikilotopic gypsum, patchily replaced by microgranular dolomite.

Interpretation. Sedimentary structures, grain shape, and sorting suggest that the sands are of aeolian origin that accumulated not as dunes, but as relatively low relief sand sheets (Hunter, 1977; Nielson and Kocurek, 1987). Invasion of the sands by saline groundwaters was probably responsible for the gypsum cement (Fryberger et al., 1983; Wasson, 1983), while the reddened and disrupted top of the unit seems to be of pedogenic origin and probably reflects partial calcretization, suggesting stabilization of the sand surface by vegetation.

3. Carbonate sediments. One of the most characteristic features of the Somerset MMG, in common with its equivalent along the northern margin of the Bristol Channel basin in South Wales (Tucker, 1978; Leslie et al., 1992), is the occurrence of decimeter-thick carbonate units. These are mainly white to light gray weathering and stand out sharply from the red, negative weathering mudstones with which they are interbedded (Fig. 10a). Carbonates become particularly important in the upper part of the MMG, where there are cyclic alternations of the two facies (Figs. 9, 10a).

Carbonate units are 10 to 30 cm thick. They typically have a gradational basal contact with the mudstones, in places marked by intense cracking and burrow mottling and a 1- to 4-cm-thick bleached zone at the top of the mudstone. The top of the carbonate bed may be gradational into the overlying mudstone, but it usually quite distinct and can be marked by desiccation cracks (Fig. 10b). Some cracks were subsequently infilled by red muds. Although thin, individual limestone units may have considerable laterally continuity and can be correlated between outcrops several kilometers apart (Fig. 9).

Internally the carbonates display a variety of structures and textures, most of which are absent in the enveloping mudstones. Primary sedimentary structures can be well preserved. They are commonly small in scale, typically bundles of millimeter-scale parallel laminae, some of which are graded, or sets of wave or current ripples (Fig. 10b). Shallow scours, many with a complex fill of alternating intraformational clast accumulations and draped sediment, are also present in some units (Fig. 10c). Lamination may be disrupted by desiccation cracks at several levels within the units, as well as at the top. Spar-filled fenestrae can also be associated with these desiccated horizons. Bioturbation is present, but is not normally as strongly developed as in the thin sandstones mentioned above. Several carbonate units display an intensely brecciated fabric (Fig. 11d). The breccias are composed of angular clasts derived from the limestone itself and show little evidence of transport, although some fragments display signs of plastic deformation. Locally, all gradients may be observed between undisturbed, well-laminated sediment and its brecciated equivalent. Thin sections of the brecciated beds show that this fabric extends down to the submillimeter scale. This lithofacies strongly resembles the breccia fabric described by Smoot and Olsen (1988) from the continental deposits of the Triassic-Jurassic Newark Supergroup.

Carbonate units are dominantly micrites or microspars. Both calcitic and dolomitic varieties are present, although the latter predominate and most calcitic rocks have at least minor developments of diagenetic dolomite. Varying amounts of silt-size clastic grains occur in all units. Also present in a few of the carbonate beds are grainstone or packstone laminae, the dominant grain type being subspherical, irregularly laminated ooids (Fig. 13c, d). In general, the sediments are unfossiliferous, although finely comminuted plant fragments occur at a few horizons, usually in association with clastic debris. Also significant is the presence of scattered *Botryococcus* colonies (Fig. 12b), apparently the first record of this alga from the British Norian. The unit in which *Botroycoccus* is most abundant is a dark limestone, relatively rich in terrestrial plant debris, which can be recognized at all three of the major west Somerset Norian exposures (Fig. 9).

Calcitic flakes, cylinders, and spheroids (Fig. 12c) form millimeter-thick packstone laminae in some of the calcitic carbonates. The spheroids are particularly striking. They are about 0.4 to 0.5 mm in diameter and have a double wall structure. Although now recrystallized and usually at least partially micritized, these grains nevertheless display a striking resemblance to charophyte gyrogonites, particularly when viewed under UV (Fig. 12d). The laminae in which these bodies occur are evidently predominantly composed of reworked charophyte debris. Like *Botryococcus,* this is the first record of charyphytes from the British Triassic, although they are known from time-equivalent rocks in Denmark (Bertelsen, 1980).

Several units, which in plane-polarized light appear to be relatively homogeneous carbonate micrites or microspars with a sprinkling of clastic grains, are seen under UV illumination to have a peloidal texture. Individual peloids are 0.05 to 0.2 mm in diameter and typically ellipsoidal in shape. Most have no internal fabric; a few show signs of an accretionary origin. In contrast to peloids in the mudstones, these are composed of micritic carbonate, or mixtures of carbonate and fine silt- to mud-grade clastic material. Associated coarser clastic grains are of similar size to the peloids, so these superficially muddy sediments are in fact clast-supported aggregates of silt-sized carbonate peloids and clastic grains cemented by microsparry calcite.

Interpretation. The well-preserved primary sedimentary structures indicate that the carbonate units are principally of aquatic origin and accumulated in relatively shallow water subject to periodic wave and current activity. Deposition probably occurred in shallow, ephemeral or semipermanent lakes. The graded units would have originated from density currents, perhaps initiated by flood inflow to the lakes. Micritic peloids seen in some of the MMG carbonates were probably formed as aeolian clay pellets (cf. Bowler, 1973; Magee, 1991) rather than as fecal pellets or from fluvial activity. A fecal origin is discarded principally because of the variations in size and shape of the peloids and the absence of any evidence for possible pellet-forming organisms. The shape and composition of the peloids, plus indications of possible accretion on the outer surface of some grains are inconsistent with an origin as soil aggregates of the sort that are so widespread in the low-gradient fluvial systems of the Australian interior. The occurrence of peloids at the base of thin graded units within the carbonates suggests that at least some of these grains may have been sorted and transported, perhaps due to flooding and erosion of exposed and desiccated mud flats or reworking of earlier-formed peloid-bearing deposits (cf. Bowler and Teller, 1986). Another similarity with Australian lakes is provided by laminae rich in charophyte fragments, formed by the grounding of drifted plant debris, and rare ooid packstones, which occur along the shallow, agitated margins of some of the larger lakes (De Deckker, 1988).

Small channels, fenestral fabrics, and abundant desiccation features indicate that the MMG lakes suffered frequent oscillations in level, the exposed sediments undergoing numerous wetting and drying cycles which led to the development of the disrupted and brecciated fabrics (Freytet and Plaziat, 1982; Smoot and Olsen, 1988). Charophyte and *Botryococcus* occurrences indicate that the lakes had fresh to brackish salinities (Burne et al., 1980), but De Deccker (1988) suggests that accumulations of ooids and charophyte debris like those seen in some of the MMG carbonates are most typical of relatively deep lake conditions, when salinities are at a minimum. The widespread occurrence of dolomite in these carbonates is also consistent with sedimentation in a lacustrine environment. It is becoming increasingly apparent that dolomite of primary or very early diagenetic (including possibly pedogenic) origin is relatively common in the muds of shallow, brackish to saline lakes in Australia (Bowler and Teller, 1986; De Deckker and Last, 1989; Rosen and Coshell, 1992) and elsewhere (Kelts and Shahrabi, 1986; Renaut and Long, 1989; Last, 1990); the ubiquitous dolomite rhombs of the Triassic deposits probably have a similar origin.

Desiccated tops to the carbonate units and cyclic alternations between these and the red mudstone suggests that climatic change may have exerted a major control on lacustrine sedimentation in the MMG, just as it has in Australia. It seems reasonable to infer that the intervening clastic mudstones accumulating during lowstands of the lakes and are thus at least in part of playa origin, particularly those intervals which contain traces of evaporites (see below). Because of prolonged periods of exposure, these parts of the lacustrine sequence were also considerably affected by pedogenic processes. Intense burrow mottling at the transition from mudstone to carbonate probably reflects the onset of wetter conditions that allowed colonization of the previously dry sediments by a relatively dense flora and fauna during and immediately prior to flooding of the depressions that eventually held standing bodies of water. The clastic-dominated units therefore represent periods of relative aridity, while the carbonates accumulated during more humid climatic intervals when deeper and more permanent waters occupied the basins. Lateral continuity of some carbonate beds over distances of several kilometers supports the notion that the sediments accumulated in relatively extensive water bodies. A similar facies association is apparent in Lake Urmia, Iran, where Holocene carbonates overlie dry, dolomitic red clastic mudstones deposited as fluvial and playa sediments during a late Pleistocene dry interval (Kelts and Shahrabi, 1986; K. Kelts, personal communication, 1991).

4. Evaporites. The dominant evaporite mineral is gypsum, which occurs at distinct horizons, mainly as nodular aggregates as much as 30 cm in diameter with a chicken-wire texture. Calcite nodules, or calcite-lined voids of similar shape to some of the smaller nodules, are also present. Elsewhere in the region these have been demonstrated to represent replaced gypsum (or anhydrite) nodules (Tucker, 1976; Berge, 1985). Veins and sheets of fibrous gypsum are also prominent at some horizons (Curtis, 1982; Wright et al., 1988). Less prominent than these sulphate accumulations are red mudstone and siltstone horizons containing abundant small (0.1 to 0.25 mm longest dimension), discoidal gypsum crystals, which occur with random orientation dispersed through the clastic matrix (Fig. 14d).

Celestite is present locally as simple nodules as much as 10 cm in diameter, which tend to be concentrated at distinct horizons (Fig. 9). Solution voids of similar size and shape to the celestite nodules suggests that this mineral may formerly have been more widespread. Substantial accumulations of celestite occur in the MMG elsewhere in southern England, some of which evidently formed by early diagenetic replacement of gypsum nodules (Thomas, 1973; Nickless et al., 1976; Curtis, 1982). The mineral is most common in marginal areas of the basin, tending to occur preferentially in a zone parallel to the basin perimeter (Nickless et al., 1976).

No halite is preserved at outcrop, but indications of its former presence are provided by rare hopper crystal casts in the mudstones. Where such casts occur, the enclosing mudstones or siltstones may have a churned texture, but the sediments may also be remarkable for the presence of a fine millimeter-scale lamination—the only horizons within the mudstone facies where primary sedimentary structures are well preserved.

Interpretation. Curtis (1982) found that gypsum occurrences in the MMG northwest of Bristol were closely associated with horizons showing clear evidence of prolonged exposure. Recently, Wright et al. (1988) have demonstrated that some of the nodular and sheet gypsum in the MMG of Somerset is related to palaeosol horizons and thus probably developed in saline soils. We endorse this interpretation and will not discuss it further, except to note that gypsiferous soils are a common feature of Quaternary sequences from arid central Australia (Bowler and Teller, 1986; Chen et al., 1991; Magee, 1991) and elsewhere (Watson, 1988, 1989). The soils formed under arid climatic conditions as a result of prolonged exposure and desiccation of muddy fluvial and playa sediments.

Horizons in the MMG rich in tiny discoidal crystals are strikingly similar to occurrences of displacive gypsum described from the margins of Australian salt lakes (cf. Fig. 14d with Bowler and Teller, 1986, Fig. 9), where they form by precipitation within the sediment at the capillary fringe of the groundwater table (Teller et al., 1982; Magee, 1991). The presence of celestite is also consistent with a lacustrine or peri-lacustrine origin for the MMG evaporites. Celestite is relatively common in the marginal deposits of some Australian salt lakes (Magee, 1991) and also occurs at the transitional zone between playa and saline mud-flat sediments of certain Californian playas (Rosen, 1991). In these settings, early diagenetic precipitation from saline groundwaters seems the most likely mode of formation for the mineral. A nodular, locally replacive habit indicates a similar origin for the MMG celestite. Several of these celestite horizons are not obviously associated with playa or lacustrine deposits (Fig. 9), suggesting that some celestite precipitation may have occurred from groundwaters moving into the basin from the surrounding uplands (cf. Rosen, 1991), rather than subsurface brines around the margins of saline lakes.

DISCUSSION

Accumulation of the MMG sequence in west Somerset occurred in continental, subaqueous and subaerial environments. Many deposits carry clear indications of prolonged periods of exposure, when they were desiccated and greatly affected by pedogenic processes. Evaporite precipitation in association with some of the intervals of soil formation suggests periods of elevated groundwater salinities, but relatively thick stratigraphic sections carrying evidence of intense bioturbation indicate that less extreme soil water conditions were common and perhaps predominant. Our discoveries of *Botryococcus,* charophytes, and ooids in some of the carbonate units confirm the periodic existence of standing bodies of water with fresh to brackish salinities, but the lakes in which these accumulated were strongly influenced by changes in the regional water balance, presumably due to oscillations between arid and wetter climatic intervals. Some aspects of the Mercia Mudstone Group still require explanation, however:

1. The MMG is dominated by fine-grained clastic sediments. In present-day Australia, although source areas are comparable and the flood-plain muds strikingly similar to the red mudstones of the MMG, major ephemeral rivers penetrate considerable distances into the interior, carrying sand as far as the northern playa of Lake Eyre (Dulhunty, 1982). Channels are typically marked by ribbons of sand within the flood-plain deposits (Rust, 1981; Hanson et al., 1988). While we cannot entirely rule out the possibility of a significant contribution to the MMG from fluvial systems of this sort, the complete absence of typically fluvial features within the main body of the mudstones makes us suspect that rivers were not the only source of fines to the basin. We have already noted the importance of aeolian processes in distributing fine-grained sediments in Australia. Aeolian dust was probably of similar or greater importance during accumulation of the MMG. Playa surfaces provide a particularly rich source of silt (Dare-Edwards, 1984; Young and Evans, 1986) and McTainsh (1989, p. 238) has proposed that terminal arid basins with humid headwaters provide the ideal combination of conditions to function as prolific source areas for wind-blown dust. The vast desiccated or saline mud surfaces that cover so much of central Australia are subjected to intense deflation. Areas to the east, downwind of this dust storm region, are thus well placed to receive large volumes of aeolian dust (McTainsh, 1985). Palaeogeographic considerations suggest that basins within which the MMG facies accumulated were also favorably located with respect to potential dust sources. The MMG basins lay at a palaeolatitude of about 20°N, at the margin of the subtropical dry zone and within the southwesterly flowing trade wind belt (Parrish and Curtis, 1982; Parrish et al., 1982, 1986). Upwind were the arid interior basin of the south-central North Sea region, a huge saline lake–sabkha complex, and beyond that a belt of mixed fluvial and aeolian sediments sourced by the Scandinavian uplands (Ziegler, 1990, Fig. 24). Deflation of surface deposits in this basin could have produced large quantities of wind-blown dust for transport out of the basin by the southwest trade winds. These fines would subsequently have been deposited as loess over the British Isles (Fig. 15). Local mud flats, alluvial fans, and upland soils would have provided additional sources of wind-blown material. From sedimentological and palaeogeographic viewpoints, we would therefore concur with those earlier workers (see above) who have suggested the MMG includes a major sediment component that is of distant origin and accumulated as loess. Vegetation trapped this material, stabilized the surface against deflation, and allowed soil organisms and pedogenic processes to rework the sediments, thoroughly mixing the aeolian and locally derived sediment components.

2. The other major question concerns the source of the dissolved salts from which the evaporites precipitated. Were these marine or continental waters, or a mixture of the two? The widespread occurrence of a sulphate/chloride mineral assemblage in the Somerset MMG, particularly the thick sequences known from the subsurface, has prompted earlier workers to suggest a marine origin for the evaporites (Evans

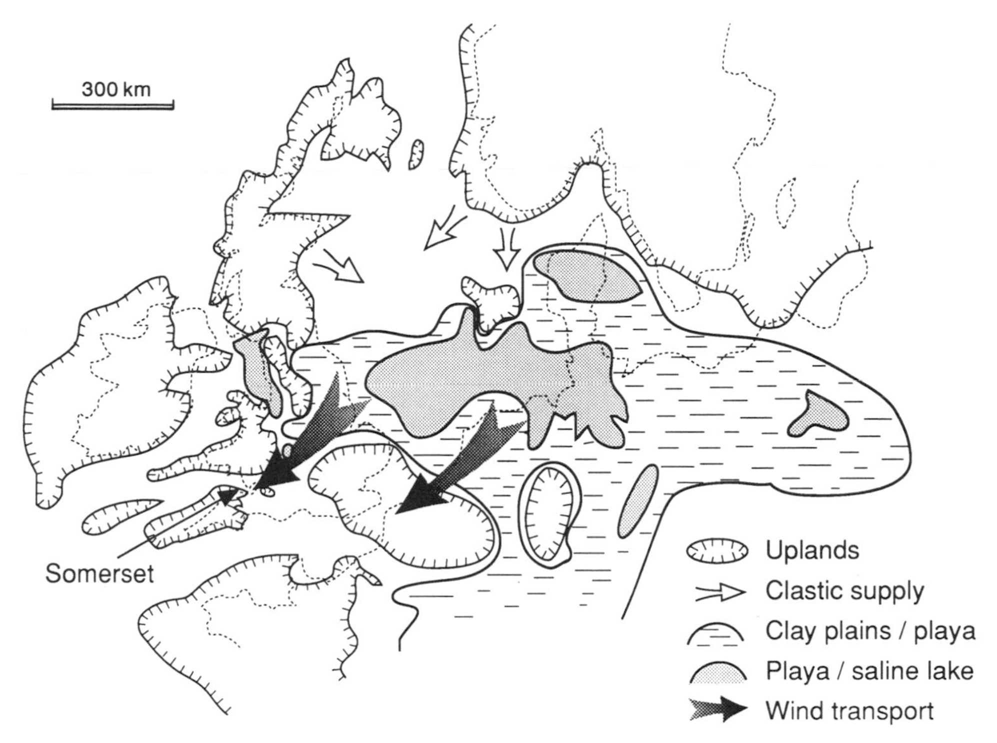

Figure 15. Late Triassic palaeogeography of northwest Europe, showing location of North Sea–Danish-German internal drainage system and probable principal transport direction for wind-blown dust deflated from playa flats. Map based on Ziegler (1990, Fig. 24); wind directions from Parrish and Curtis (1982).

et al., 1968; Warrington, 1974; Arthurton, 1980; Holloway, 1985), perhaps via marine incursions into the basin. This has in turn led to suggestions that marine processes have had an important influence upon sedimentation in general (e.g. Wills, 1970). The evaporite mineral assemblage from the Quaternary of central Australia also has a distinctly marine character, but derivation of these salts directly from seawater is out of the question. A major aerosol contribution of marine sulphate has recently been demonstrated for parts of Australia (Chivas et al., 1991) and marine aerosols also seem capable of supplying all the chloride present in some Californian interior basins (Rosen, 1991). A direct marine origin is probably not required for the MMG evaporites, either. Hardie (1984, 1991) has noted that thick sequences of bedded, sulphate-poor halite, such as occur in the central part of the Somerset basin (Whittaker, 1980), are more characteristic of nonmarine than marine evaporite basins. Some salts may have been supplied by aerosols—the sea was probably located only a few hundred kilometers to the west and south (Ziegler, 1990)—and an additional important source of sulphate must have been the weathering of sedimentary sulphide minerals in the thick upper Palaeozoic marine shales and carbonate mudstones that surround the basin. Stable isotopic studies of MMG gypsum from the South Wales part of Bristol Channel Basin have indicated a dominantly continental source for the sulphate (Taylor, 1983); variable mixtures of aerosol and weathering sulphate can probably explain some of the sulphur isotopic variations recorded by Taylor. The source of strontium for the MMG celestite deposits has also been a subject of considerable debate. Strontium isotope data (Wood and Shaw, 1976) indicate that the celestite is enriched in radiogenic Sr, which precludes direct precipitation from contemporary seawater or reworking of earlier marine evaporites or carbonates. Other possibilities are derivation from hydrothermal springs (there is widespread evidence for local hydrothermal activity in the early Mesozoic basins of southwest England) or *via* contemporary weathering of exposed strontium-rich barite veins that are relatively common in the exposed Palaeozoic basement (Thomas, 1973; Nickless et al., 1976).

DEPOSITIONAL MODEL FOR THE MERCIA MUDSTONE GROUP

Our depositional model for the MMG is summarized in Figure 16. Sediment types, sedimentary processes, and chemical sediment composition are all consistent with a wholly continental origin for the MMG of west Somerset. There were evidently marked variations in climate, which exerted a major control upon sedimentation. Cyclic alternations of lacustrine carbonates and red terrestrial mudstones in the upper part of the MMG (Figs. 9, 10a) raise the possibility of an astronomi-

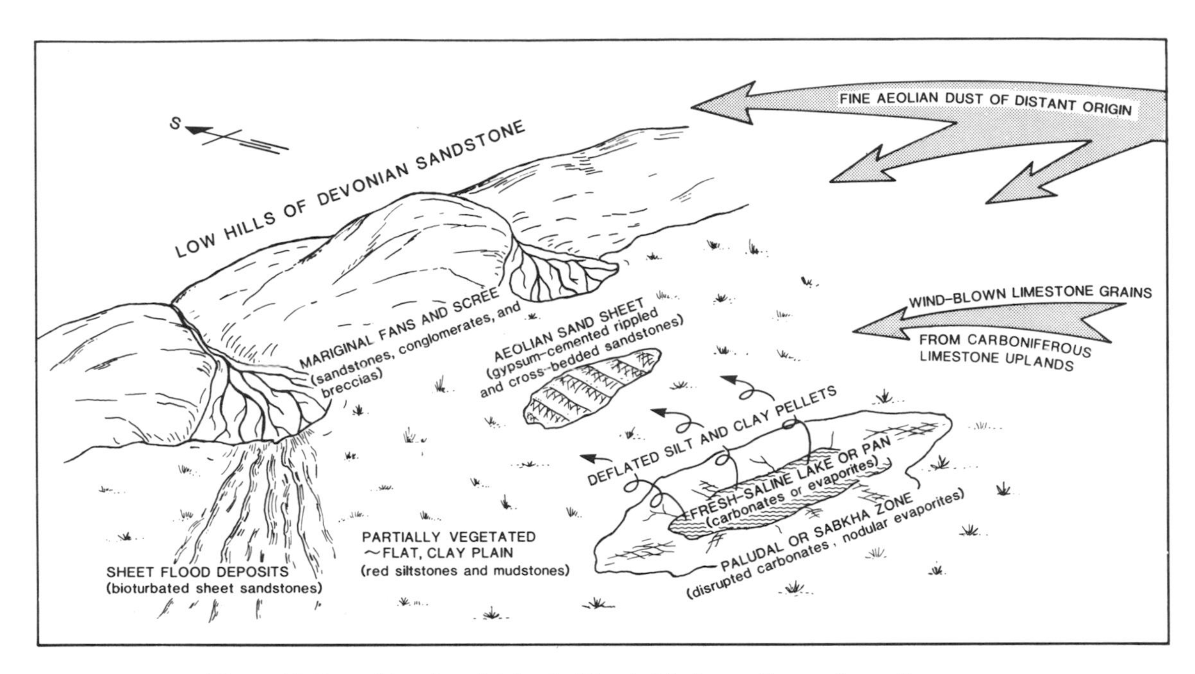

Figure 16. Depositional model for the Mercia Mudstone Group of west Somerset.

cal control on the occurrence of dry and humid periods. Stratigraphic resolution is not sufficiently refined to allow any estimate of the periodicity of climatic change during the period of MMG accumulation, but the demonstration of Milankovitch-type cycles in time-equivalent continental rocks of the Newark Supergroup, eastern United States (Olsen, 1984, 1986) strongly suggests that astronomical forcing may have been responsible for the clastic-carbonate cycles of the MMG.

One general lesson that can be drawn from this comparative study is that the massive, typically red mudstones that characterize the fill of many arid-semiarid continental basins are probably polygenetic. They may be lacustrine (including playa), fluvial, or aeolian, or more probably, a mixture of all three. In sequences of mixed origin we presently lack reliable criteria that can be used to distinguish between the products of these three processes. Such distinctions may be unattainable. As we have shown here, the fluvial and lacustrine environments in these low-gradient systems grade imperceptibly into each other across vague boundaries that constantly shift with variations in sediment supply and climate. Gradual trapping and accumulation of aeolian dust on the exposed surface of these sediments inevitably leads to mixing of the terrestrial and windblown components. Furthermore, the end result of prolonged periods of exposure, desiccation, bioturbation, and pedogenesis is a sequence of relatively uniform mudstones and siltstones within which many of the individual depositional units are barely distinguishable.

SUMMARY

1. The west Somerset MMG sediments accumulated in a low-relief interior basin flanked by subdued uplands of upper Palaeozoic sedimentary rocks. Away from the basin margins, the environment was principally a monotonous clay plain, which during humid periods and following rains was dotted by shallow, fresh water to saline lakes.

2. During arid and semiarid intervals, sedimentation was dominated by the accumulation of clastic muds and silts. Some of the fine-grained sediments were supplied by fluvial systems draining the adjacent uplands and accumulated from ponded waters on flood plains and in playas; during rare extreme floods sandy sediments were transported far into the basin. Periods of particularly intense aridity are marked by rare aeolian sand accumulations and the development of saline groundwater conditions. At other times the surface carried vegetation and newly deposited sediments were actively reworked by pedoturbation and soil organisms.

3. Humid intervals were marked by the widespread development of lakes, but there is no convincing evidence for the existence of a single, major waterbody occupying much of the Bristol Channel basin during the Norian (cf. Tucker, 1977, 1978). The sedimentary evidence best fits the development of small, shallow waterbodies that temporarily (10s to 1,000s of years) occupied local erosional and structural depressions. Carbonate or mixed carbonate-clastic sediments characterized the periods of most dilute water composition in these lakes. The bulk of this carbonate has presumably been derived from the weathering of older rocks exposed in the basin catchment. Adjacent areas of Palaeozoic carbonates and calcite-cemented sandstones would have ensured a plentiful supply of carbonate-rich inflow to the basin. At least some of this flow was probably derived from extensive karstic spring systems that developed in the Carboniferous Limestone (cf. Leslie et al., 1992).

4. Aeolian processes supplied significant volumes of silt and clay-sized material to the basin. Some of this was probably of distant origin, derived from arid basins located upwind (northeast) of southern Britain, but the wind probably also transported material from nearby upland soils, from the surfaces of marginal alluvial fans and muds newly exposed around the margins of shrinking lakes.

5. Apart from scattered halite pseudomorphs, there is no clear evidence at outcrop for the existence of waterlaid evaporites in the MMG, perhaps because the residual waterbodies that existed at these times were located in the topographically lowest part of the basin, offshore from the present MMG outcrop. Sulphates in the exposed parts of the MMG seem to have formed mainly by evaporation from the capillary fringe of shallow groundwater tables, some around the margins of playa lakes (this study), some within the distal deposits of low-angle alluvial fan systems (Wright, 1992). The evaporite assemblage in the exposed parts of the Somerset MMG is consistent with the analogies we have been drawing with arid Australia, where bedded halite is rare and only a few saline lakes contain bedded sulphates (M. R. Rosen, personal communication, 1991).

ACKNOWLEDGMENTS

M. R. Talbot thanks Macquarie University for the travel grant that made possible his visit to Australia. Kristine Holm acknowledges financial support for fieldwork in Somerset from Amoco Norway Inc. M.A.J. Williams thanks the Australian Research Council, Macquarie University, and Monash University for their financial support. We all thank Don Adamson for many campfire discussions on the arid zones of the world and Paul Wright and Michael Rosen for valuable reviews of an earlier version of this paper. The skills of Jane Ellingsen and Masaoki Adachi turned sketches into presentable art.

REFERENCES CITED

Arthurton, R. S., 1980, Rhythmic sedimentary sequences in the Triassic Keuper Marl (Mercia Mudstone Group) of Cheshire, northwest England: Geological Journal, v. 15, p. 43–58.

Begin, Z. B., Ehrlich, A., and Nathan, Y., 1974, Lake Lisan the Pleistocene precursor of the Dead Sea: Geological Survey of Israel Bulletin, v. 63, p. 1–30.

Berge, C., 1985, Et Sedimentologisk og Diagenetisk Studium av Bergarter Tilhørende "Marginal" Trias ved Clevedon og Portishead, Avon, England [M.Sc. thesis]: University of Bergen, 444 p.

Bertelsen, F., 1980, Lithostratigraphy and depositional history of the Danish Triassic: Danmarks Geologiske Undersøgelse, v. serie B, no. 4, p. 1–59.

Birkeland, P. W., 1984, Soils and geomorphology: Oxford, Oxford University Press, 372 p.

Blodgett, R. H., 1988, Calcareous paleosols in the Triassic Dolores Formation, southwestern Colorado, *in* Reinhardt, J., and Sigleo, W. R., eds., Paleosols and weathering through geologic time: Boulder, Colorado, Geological Society of America Special Paper 216, p. 103–121.

Bowler, J. M., 1973, Clay dunes: Their occurrence, formation and environmental significance: Earth-Science Reviews, v. 9, p. 315–338.

Bowler, J. M., 1976, Aridity in Australia: Age origins and expression in aeolian landforms and sediments: Earth-Science Reviews, v. 12, p. 279–310.

Bowler, J. M., 1981, Australian salt lakes: A palaeohydrological approach: Hydrobiologia, v. 82, p. 431–444.

Bowler, J. M., 1986, Spatial variability and hydrologic evolution of Australian lake basins: Analogue for Pleistocene hydrologic change and evaporite formation: Palaeogeography, Palaeoclimatology, Palaeoecology, v. 54, p. 21–41.

Bowler, J. M., and Teller, J. T., 1986, Quaternary evaporites and hydrological changes, Lake Tyrell, north-west Victoria: Australian Journal of Earth Sciences, v. 33, p. 43–63.

Brewer, R., 1976, Fabric and mineral analysis of soils: New York, Krieger, 482 p.

Brookfield, M., 1970, Dune trends and wind regime in central Australia: Zeitschrift für Geomorphologie Supplementband, v. 10, p. 121–153.

Burne, R. V., Bauld, J., and De Deckker, P., 1980, Saline lake charophytes and their geological significance: Journal of Sedimentary Petrology, v. 50, p. 281–293.

Butler, B. E., 1956, Parna—An aeolian clay: Australian Journal of Science, v. 18, p. 145–151.

Chadwick, R. A., 1986, Extension tectonics in the Wessex Basin, southern England: Geological Society of London Journal, v. 143, p. 465–488.

Chafetz, H. S., Rush, P. F., and Utech, N. M., 1991, Microenvironmental controls on mineralogy and habit of $CaCO_3$ precipitates: An example from an active travertine system: Sedimentology, v. 38, p. 107–126.

Chen, X. Y., Bowler, J. M., and Magee, J. W., 1991, Aeolian landscapes in central Australia: Gypsiferous and quartz dune environments from Lake Amadeus: Sedimentology, v. 38, p. 519–538.

Chivas, A. R., Andrew, A. S., Lyons, W. B., Bird, M. I., and Donnelly, T. H., 1991, Isotopic constraints on the origin of salts in Australian playas. 1. Sulphur: Palaeogeography, Palaeoclimatology, Palaeoecology, v. 84, p. 309–332.

Curtis, M. T., 1982, Playa cycles in the Mercia Mudstone (Keuper Marl) of Aust Cliff, Avon: Proceedings, Bristol Naturalists' Society, v. 42, p. 13–22.

Dare-Edwards, A. J., 1984, Aeolian clay deposits of south-eastern Australia: Parna or loessic clay?: Transactions of the Institute of British Geographers New Series, v. 9, p. 337–344.

De Deckker, P., 1983, Australian salt lakes: Their history, chemistry, and biota—A review: Hydrobiologia, v. 105, p. 231–244.

De Deckker, P., 1988, Biological and sedimentary facies of Australian salt lakes: Palaeogeography, Palaeoclimatology, Palaeoecology, v. 62, p. 237–270.

De Deckker, P., and Last, W. M., 1989, Modern, non-marine dolomite in evaporitic playas of western Victoria, Australia: Sedimentary Geology, v. 64, p. 223–238.

Demicco, R. V., and Gierlowski-Kordesch, E. G., 1986, Facies sequences of a semi-arid closed basin: The Lower Jurassic East Berlin Formation of the Hartford basin, New England, USA: Sedimentology, v. 33, p. 107–118.

Dravis, J. J., and Yurewicz, D. A., 1985, Enhanced carbonate petrography using fluorescence microscopy: Journal of Sedimentary Petrology, v. 55, p. 795–804.

Dulhunty, J. A., 1982, Holocene sedimentary environments in Lake Eyre, South Australia: Journal of the Geological Society of Australia, v. 29, p. 437–442.

Dumbleton, M. J., and West, G., 1966, Studies of the Keuper Marl: Mineralogy: Ministry of Transport Road Research Laboratory Report, v. 40, p. 1–25.

Evans, W. B., Wilson, A. A., Taylor, B. J., and Price, D., 1968, Geology of the country around Macclesfield, Congleton, Crewe and Middlewich: London, Geological Survey of Great Britain, Memoir, p. 1–328.

Frey, R. W., Pemberton, S. G., and Fagerstrom, J. A., 1984, Morphological, ethological, and environmental significance of the ichnogenera Scoyenia and Ancorichnus: Journal of Paleontology, v. 58, p. 511–528.

Freytet, P., and Plaziat, J. C., 1982, Continental carbonate sedimentation and pedogenesis—Late Cretaceous and early Tertiary of southern France: Contributions to Sedimentology, v. 12, p. 1–213.

Fryberger, S. G., Al-Sari, A. M., and Clisham, T. J., 1983, Eolian dune, interdune, sand sheet and siliclastic sabkha sediments of an offshore prograding sand sea, Dharan area, Saudi Arabia: American Association of Petroleum Geologists Bulletin, v. 67, p. 280–312.

Gore, P.J.W., 1988, Late Triassic and Early Jurassic lacustrine sedimentation in the Culpeper Basin, Virginia, *in* Manspeizer, W., ed., Triassic-Jurassic rifting: Continental breakup and the origin of the Atlantic Ocean and passive margins: Amsterdam, Elsevier, p. 369–400.

Green, G. W., and Welch, F.B.A., 1965, Geology of the country around Wells and Cheddar: London, Geological Survey of Great Britain Memoir, 225 p.

Gustavson, T. C., 1991, Buried vertisols in a lacustrine facies of the Pliocene Fort Hancock Formation, Hueco Bolson, west Texas and Chihuahua, Mexico: Geological Society of America Bulletin, v. 103, p. 448–460.

Hardie, L. A., 1984, Evaporites: marine or non-marine? American Journal of Science, v. 284, p. 193–240.

Hardie, L. A., 1991, On the significance of evaporites: Annual Review of Earth and Planetary Sciences, v. 19, p. 131–168.

Hardie, L. A., Smoot, J. P., and Eugster, H. P., 1978, Saline lakes and their deposits: A sedimentological approach, *in* Matter, A., and Tucker, M. E., Eds., Modern and Ancient Lake Sediments: International Association of Sedimentologists Special Publication 2, p. 7–41.

Hay, R. L., and Wiggins, B., 1980, Pellets, ooids, sepiolite and silica in three calcretes of the southwestern United States: Sedimentology, v. 27, p. 559–576.

Holloway, S., 1985, Triassic: Mercia Mudstone and Penarth groups, *in* Whittaker, A., ed., Atlas of onshore sedimentary basins in England and Wales: Post-Carboniferous tectonics and stratigraphy: Glasgow, Blackie, p. 34–36.

Hunter, R. E., 1977, Basic types of stratification in small eolian dunes: Sedimentology, v. 24, p. 361–387.

Jeans, C. V., 1978, The origin of the Triassic clay assemblages of Europe with special reference to the Keuper Marl and Rhaetic of parts of England: Philosophical Transactions of the Royal Society, v. A289, p. 551–636.

Kamerling, P., 1979, The geology and hydrocarbon habitat of the Bristol Channel basin: Journal of Petroleum Geology, v. 2, p. 75–93.

Kelts, K., and Shahrabi, M., 1986, Holocene sedimentology of hypersaline Lake Urmia, northwestern Iran: Palaeogeography, Palaeoclimatology, Palaeoecology, v. 54, p. 105–130.

Kendall, A. C., 1984, Evaporites, *in* Walker, R. G., ed., Facies models: Toronto, Geoscience Canada Reprint Series 1, p. 259–296.

Last, W. M., 1990, Lacustrine dolomite—An overview of modern, Holocene and Pleistocene occurrences: Earth Science Reviews, v. 27, p. 221–263.

Leslie, A. B., Tucker, M. E., and Spiro, B., 1992, A sedimentological and stable isotopic study of travertines and associated sediments within the Upper Triassic lacustrine limestones, South Wales, U.K.: Sedimentology, v. 39, p. 613–630.

Lomas, J., 1907, Desert conditions and the origin of the British Trias: Geological Magazine, v. 4, p. 511–514 and 554–563.

Magee, J. W., 1991, Late Quaternary lacustrine, groundwater, aeolian and pedogenic gypsum in the Prungle Lakes, southeastern Australia: Palaeogeography, Palaeoclimatology, Palaeoecology, v. 84, p. 3–42.

McTainsh, G. H., 1985, Dust processes in Australia and West Africa: A comparison: Search, v. 16, p. 104–106.

McTainsh, G. H., 1989, Quaternary aeolian dust processes and sediments in the Australian region: Quaternary Science Reviews, v. 8, p. 235–253.

Nanson, G. C., Rust, B. R., and Taylor, G., 1986, Coexistent mud braids and anastomosing channels in an arid-zone river: Cooper Creek, central Australia: Geology, v. 14, p. 175–178.

Nanson, G. C., Young, R. W., Price, D. M., and Rust, B. R., 1988, Stratigraphy, sedimentology and late Quaternary chronology of the channel country of western Queensland, *in* Warner, R. F., ed., Fluvial geomorphology of Australia: Australia, Academic Press, p. 151–175.

Nickless, E.F.P., Booth, S. J., and Mosley, P. N., 1976, The celestite resources of the area south-east of Bristol, with notes on occurrences north and south of the Mendip Hills and in the Vale of Glamorgan: Mineral Assessment Report Institute of Geological Sciences, v. 25, p. 1–2.

Nielson, J., and Kocurek, G., 1987, Surface processes, deposits, and development of star dunes: Dumont dune field, California: Geological Society of America Bulletin, v. 99, p. 177–186.

Olsen, P. E., 1984, Periodicity of lake-level cycles in the Late Triassic Lockatong Formation of the Newark Basin (Newark Supergroup, New Jersey and Pennsylvania), *in* Berger, A. L., ed., Milankovitch and Climate, Part 1: Dordrecht, Reidel, p. 129–146.

Olsen, P. E., 1986, A 40-million-year lake record of early Mesozoic orbital climatic forcing: Science, v. 234, p. 842–848.

Olsen, P. E., 1988, Paleontology and paleoecology of the Newark Supergroup (early Mesozoic, eastern North America), *in* Manspeizer, W., ed., Triassic-Jurassic rifting: Amsterdam, Elsevier, p. 185–230.

Parrish, J. M., Parrish, J. T., and Ziegler, A. M., 1986, Permian-Triassic paleogeography and paleoclimatology and implications for Theraspid distribution, *in* Hotton, N., MacLean, P. D., Roth, J. J., and Roth, E. C., eds., The ecology and biology of mammal-like reptiles: Washington, D.C., Smithsonian Institution, p. 109–131.

Parrish, J. T., and Curtis, R. L., 1982, Atmospheric circulation, upwelling, and organic-rich rocks in the Mesozoic and Cenozoic eras: Palaeogeography, Palaeoclimatology, Palaeoecology, v. 40, p. 31–66.

Parrish, J. T., Ziegler, A. M., and Scotese, C. R., 1982, Rainfall patterns and the distribution of coals and evaporites in the Mesozoic and Cenozoic: Palaeogeography, Palaeoclimatology, Palaeoecology, v. 40, p. 67–101.

Petit-Maire, N., ed., 1982, Le Shati, lac pléistocène du Fezzan: Marseille, Centre National de la Recherche Scientifique, 118 p.

Petit-Maire, N., 1986, Palaeoclimates in the Sahara of Mali: Episodes, v. 9, p. 7–16.

Price, W. A., 1963, Physicochemical and environmental factors in clay dune genesis: Journal of Sedimentary Petrology, v. 33, p. 766–778.

Renaut, R. W., and Long, P. R., 1989, Sedimentology of the saline lakes of the Cariboo Plateau, interior British Columbia, Canada: Sedimentary Geology, v. 64, p. 239–264.

Retallack, G. J., 1983, Late Eocene and Oligocene paleosols from Badlands National Park, South Dakota: Geological Society of America Special Paper 193, 82 p.

Retallack, G. J., 1988, Field recognition of paleosols, *in* Reinhardt, J., and Sigleo, W. R., eds., Paleosols and weathering through geologic time: Boulder, Colorado, Geological Society of America Special Paper 216, p. 1–20.

Ritchie, J. C., Eyles, C. H., and Haynes, C. V., 1985, Sediment and pollen evidence for an early to mid-Holocene humid period in the eastern Sahara: Nature, v. 314, p. 352–355.

Rosen, M. R., 1991, Sedimentologic and geochemical constraints on the evolution of Bristol Dry Lake basin, California, U.S.A.: Palaeogeography, Palaeoclimatology, Palaeoecology, v. 84, p. 229–257.

Rosen, M. R., and Coshell, L., 1992, A new location of Holocene dolomite formation, Lake Hayward, Western Australia: Sedimentology, v. 39, p. 161–166.

Rust, B. R., 1981, Sedimentation in an arid-zone anastomosing fluvial system: Cooper's Creek, central Australia: Journal of Sedimentary Petrology, v. 51, p. 745–755.

Rust, B. R., and Legun, A. S., 1983, Modern anastomosing-fluvial deposits in arid central Australia and a Carboniferous analogue in New Brunswick, Canada, *in* Collinson, J. D., and Lewin, J., eds., Modern and ancient fluvial systems: Oxford, International Association of Sedimentologists Special Publication 6, p. 385–392.

Rust, B. R., and Nanson, G. C., 1989, Bedload transport of mud as pedogenic aggregates in modern and ancient rivers: Sedimentology, v. 36, p. 291–306.

Simms, M. J., and Ruffell, A. H., 1990, Climatic and biotic change in the Late Triassic: Geological Society of London Journal, v. 147, p. 321–327.

Smoot, J. P., 1991, Sedimentary facies and depositional environments of early Mesozoic Newark Supergroup basins, eastern North America: Palaeogeography, Palaeoclimatology, Palaeoecology, v. 84, p. 369–423.

Smoot, J. P., and Olsen, P. E., 1988, Massive mudstone in basin analysis and paleoclimatic interpretation of the Newark Supergroup, *in* Manspeizer, W., ed., Triassic-Jurassic rifting: Amsterdam, Elsevier, p. 249–274.

Street-Perrott, F. A., and Roberts, N., 1983, Fluctuations in closed basin lakes as indicators of past atmospheric circulation patterns, in Street-Perrott, F. A., Beran, M., and Ratcliffe, R.A.S., eds., Variations in the global water budget: Dordrecht, Reidel, p. 331–345.

Talbot, M. R., 1984, Ultra-violet microscopy in the study of carbonate rocks, *in* Bathurst, R.G.C., ed., 7th Meeting of Carbonate Sedimentologists: Liverpool, University of Liverpool, p. 64.

Talbot, M. R., 1988, The origins of lacustrine oil source rocks: Evidence from the lakes of tropical Africa, *in* Fleet, A. J., Kelts, K., and Talbot, M. R., eds., Lacustrine petroleum source rocks: London, Geological Society of London Special Publication 40, p. 29–43.

Taylor, B. J., Price, R. H., and Trotter, F. M., 1963, Geology of the country around Stockport and Knutsford: London, Geological Survey of Great Britain, Memoir, 183 p.

Taylor, S. R., 1983, A stable isotope study of the Mercia Mudstones (Keuper Marl) and associated sulphate horizons in the English Midlands: Sedimentology, v. 30, p. 11–31.

Teller, J. T., and Last, W. M., 1990, Paleohydrological indicators in playas and salt lakes, with examples from Canada, Australia and Africa: Palaeogeography, Palaeoclimatology, Palaeocology, v. 76, p. 215–240.

Teller, J. T., Bowler, J. M., and Macumber, P. G., 1982, Modern sedimentation and hydrology in Lake Tyrrell, Victoria: Geological Society of Australia Journal, v. 29, p. 159–175.

Thomas, I. A., 1973, Celestite: Institute of Geological Sciences Mineral Resources Division Mineral Dossier, v. 6, p. 1–26.

Tiercelin, J. J., Renaut, R. W., Delibrias, G., Le Fournier, J., and Bieda, S., 1981, Late Pleistocene and Holocene lake level fluctuations in the Lake Bogoria basin, northern Kenya Rift Valley: Palaeoecology of Africa, v. 13, p. 105–120.

Torgersen, T., De Deckker, P., Chivas, A. R., and Bowler, J. M., 1986, Salt lakes: A discussion of processes influencing palaeoenvironmental interpretation and recommendations for future study: Palaeogeography, Palaeoclimatology, Palaeoecology, v. 54, p. 7–19.

Tucker, M. E., 1976, Quartz-replaced anhydrite nodules ("Bristol Diamonds") from the Triassic of the Bristol district: Geological Magazine, v. 113, p. 569–574.

Tucker, M. E., 1977, The marginal Triassic deposits of South Wales: Continental facies and palaeogeography: Geological Journal, v. 12, p. 169–188.

Tucker, M. E., 1978, Triassic lacustrine sediments from South Wales: Shorezone clastics, evaporites and carbonates, *in* Matter, A., and Tucker, M. E., eds., Modern and ancient lake sediments: Oxford, International Association of Sedimentologists Special Publication 2, p. 205–224.

Valeton, I., 1978, A morphological and petrological study of the terraces around Lake Van, Turkey, *in* Degens, E. T., and Kurtman, F., eds., The geology of Lake Van: Ankara, M.T.A. Press, p. 64–80.

Van Houten, F. B., 1962, Cyclic sedimentation and the origin of analcime-rich Upper Triassic Lockatong Formation, west-central New Jersey and adjacent Pennsylvania: American Journal of Science, v. 260, p. 561–576.

Van Houten, F. B., 1964, Cyclic lacustrine sedimentation, Upper Triassic Lockatong Formation, central New Jersey and adjacent Pennsylvania: Kansas Geological Survey Bulletin, v. 169, p. 497–531.

Warren, J. K., 1982, The hydrological setting, occurrence and significance of gypsum in late Quaternary salt lakes in South Australia: Sedimentology, v. 29, p. 609–637.

Warrington, G., 1974, Les évaporites du Trias britannique: Bulletin de la Societé Géologique de France, v. 16, p. 708–723.

Warrington, G., and 8 others, 1980, A correlation of Triassic rocks in the British Isles: Geological Society of London Special Report, v. 13, p. 1–78.

Wasson, R. J., 1976, Holocene aeolian landforms in the Belarabon area, S.W. of Cobar, N.S.W.: Royal Society of New South Wales Journal and Proceedings, v. 109, p. 91–101.

Wasson, R. J., 1983, The Cainozoic history of the Strzelecki and Simpson dunefields (Australia), and the origin of desert dunes: Zeitschrift für Geomorphologie Supplementband, v. 45, p. 85–115.

Wasson, R. J., 1986, Geomorphology and Quaternary history of the Australian continental dunefields: Geographical Review of Japan, v. 59, p. 55–67.

Watson, A., 1988, Desert gypsum crusts as palaeoenvironmental indicators: A micropetrographic study of crusts from southern Tunisia and the central Namib Desert: Journal of Arid Environments, v. 15, p. 19–42.

Watson, A., 1989, Desert crust and rock varnish, *in* Thomas, D.S.G., ed, Arid zone geomorphology: London, Belhaven Halsted, p. 25–55.

Whittaker, A., 1980, Triassic salt deposits in southern England: Fifth International Symposium on Salt, v. 1, p. 175–179.

Whittaker, A., and Green, G. W., eds., 1983, Geology of the country around Western-super-Mare: London, Institute of Geological Sciences, 147 p.

Williams, M.A.J., De Deckker, P., Adamson, D. A., and Talbot, M. R., 1991, Episodic fluviatile, lacustrine and aeolian sedimentation in a late Quaternary desert margin system, central western New South Wales, *in* Williams, M.A.J., De Deckker, P., and Kershaw, A. P., eds., The Cainozoic in Australia: A re-appraisal of the evidence: Sydney, Geological Society of Australia Special Publication 8, p. 258–287.

Wills, L. J., 1970, The Triassic succession in the central Midlands: Geological Society of London Journal, v. 126, p. 225–283.

Wills, L. J., 1976, The Trias of Worcestershire and Warwickshire: Institute of Geological Sciences Report, v. 76/2, p. 1–211.

Wood, M. W., and Shaw, H. F., 1976, The geochemistry of celestites from the Yate area near Bristol (U.K.): Chemical Geology, v. 17, p. 179–193.

Woodyer, K. D., Taylor, G., and Crook, K.A.W., 1979, Depositional processes along a very low-gradient, suspended-load stream: The Barwon River, New South Wales: Sedimentary Geology, v. 22, p. 97–120.

Wright, V. P., 1992, Paleopedology: Stratigraphic relationships and empirical models, *in* Martini, I. P., and Chesworth, W., eds., Weathering, Soils and paleosols: Amsterdam, Elsevier, p. 475–499.

Wright, V. P., North, C. P., Hancock, P. L., Curtis, M., and Robinson, D., 1988, Pedofacies variations across an arid alluvial basin: A case-study from the Upper Triassic of SW Britain: Ninth International Association of Sedimentologists Regional Meeting of Sedimentology, Abstracts, Leuven, p. 227–228.

Young, J. A., and Evans, R. A., 1986, Erosion and deposition of fine sediment from playas: Journal of Arid Environments, v. 10, p. 103–115.

Ziegler, P. A., 1990, Geological Atlas of western and central Europe: The Hague, Shell Internationale Petroleum Maatschappij B.V., 239 p.

MANUSCRIPT ACCEPTED BY THE SOCIETY JULY 2, 1993

Printed in U.S.A.

Typeset by WESType Publishing Services, Inc., Boulder, Colorado
Printed in U.S.A. by Johnson Printing, Boulder, Colorado

The Geological Society of America
3300 Penrose Place • P.O. Box 9140 • Boulder, Colorado 80301

Contents

ISBN 0-8137-2289-6